IET CONTROL ENGINEERING SERIES 8

Series Editors: G.A. Montgomerie
Prof. H. Nicholson
Prof. B.H. Swanick

T0258013

A History of
Control Engineering
1800–1930

Other volumes in this series:

A History of
Control Engineering
1800–1930

S. Bennett

The Institution of Engineering and Technology

Published by The Institution of Engineering and Technology, London, United Kingdom

First edition © 1979 Peter Peregrinus Ltd
Reprint with new cover @ 2008 The Institution of Engineering and Technology

First published 1979
Paperback edition 1986
Reprinted 2008

The Institution of Engineering and Technology
Michael Faraday House
Six Hills Way, Stevenage
Herts, SG1 2AY, United Kingdom

www.theiet.org

British Library Cataloguing in Publication Data
A history of control engineering, 1800-1930
(Control engineering series no 8)
1. Feedback control systems – History
I. Title II. Series
629.8'3 09034   TJ216

ISBN (10 digit) 0 86341 047 2
ISBN (13 digit) 978-0-86341-047-5

Printed in the UK by Lightning Source UK Ltd, Milton Keynes

To M. F. W.

Contents

... if every instrument could accomplish its own work, obeying or anticipating the will of others ... if the shuttle would weave and the pick touch the lyre without a hand to guide them, chief workmen would not need servants, nor masters slaves.

Aristotle, *Politics*, Book 1, chapter 3

Preface

It has often been said that a discipline reaches maturity only when it becomes aware of its own history: if this be so, control engineering reached maturity in 1969 when Otto Mayr's book *Zur Frühgeschichte der Technischen Regelungen* was published in West Germany (an English edition of the book entitled *The origins of feedback control* appeared in the following year).* In 1969, also, an English translation of Krhamoi's book on the history of automatic control in Russia was issued and a dissertation entitled *Zur Entwicklung der Regelungstechnik* was submitted to the University of Erlangen by Klaus Rörentrop, later published under the title *Die Entwicklung der Modernen Regelungstechnik* in 1971. Prior to 1969 only two papers of importance on the history of control, those by Conway and Ramsey, had been published; however, since 1970 interest in the subject has grown and papers have been published by Mayr, Fuller, myself and others.

My interest in the history of control engineering began in about 1968 and was stimulated by the publications of Otto Mayr. In this book I have taken up the story where Mayr left off — at the invention of the governor — and have continued it to about 1930. I have attempted to cover the main practical and theoretical developments of this period, but, to keep the book to a manageable size, the history of instruments and process control has been largely excluded. During the preparation of this book I have received constant help and encouragement from Dr. Otto Mayr, Curator of Mechanical Sciences, The Smithsonian Institute, and from Dr. Tom Fuller, of the University of Cambridge; I am grateful to both and also to Professor H. Nicholson, Head of the Department of Control Engineering, University of Sheffield, for his

---

* A short bibliography of books and articles on the history of control systems is given on p. 204.

support. I am also grateful to the staff of the Applied Science Library, University of Sheffield, and to Jose Stubbs, Eileen Halse and Pauline Turner for their help in preparing the typescript. Finally, I would like to thank my wife who has taken time off from her own writing to read each draft and has improved every page she read; perhaps she will now have time to finish her work.

Stuart Bennett
University of Sheffield   February 1978

Chapter 1

Feedback: the origins
of a concept

Perhaps his most curious idea was for a contrivance by which a horse driving a mill or otherwise should bring into action an automatic goad, so that if he did not pull with a certain force it should prick him until he did.

E. A. Forward, writing in 1922 on an 18th-century engineer, Simon Goodrich

One of the most common examples of a boring fad is the predilection for the word 'feedback' which (outside its proper technical context) usually merely replaces more precise words like 'report' or 'reaction'.

S. Andreski, 1972

The word 'feedback' seems to have been used first in 1920[*] and, although initially it had to vie with the terms 'reset', used in England, and 'closed-cycle', used in the USA, its adoption in the 1920s by the communications engineers at the Bell Telephone Laboratory ensured its eventual ascendancy. As a concept it is, however, much older, although it emerged not in engineering but in political economy:

It is a curious fact that, while political economists recognise that for the proper action of the law of supply and demand there must be fluctuations, it has not generally been recognised by mechanicians in this matter of the steam engine governor. The aim of the mechanical economist, as it is that of the political economist, should be not to do away with these fluctuations altogether (for then he does away with the principles of self-regulation), but to diminish them as much as possible, still leaving them large enough to have sufficient regulating power.[1]

[*] Supplement to the Oxford English Dictionary, 1972: 'An inductive feed-back in relation to the secondary system generates local oscillations', *Wireless Age*, 1920, VIII

In political economics, the idea of a closed cycle was implicit in the writings of David Hume:

> Suppose four-fifths of all the money of GREAT BRITAIN to be annihilated in one night, and the nation reduced to the same condition, with regard to specie, as in the reigns of the HARRYS AND EDWARDS, what would be the consequence? Must not the price of all labour and commodities sink in proportion, and everything be sold as cheap as they were in those ages? What nation could then dispute with us in any foreign market, or pretend to navigate or to sell manufactures at the same price, which to us would afford sufficient profit? In how little time, therefore, must this bring back the money which we had lost, and raise us to the level of all the neighbouring nations? Where after we have arrived, we immediately lose the advantage of the cheapness of labour and commodities; and the farther flowing in of money is stopped by our fulness and repletion.[2]

while, in the *Wealth of nations,* Adam Smith provides three examples of well defined social feedback mechanisms: wages, population and general supply and demand.[3] In 1931 Kahn expressed in more formal terms the closed-cycle connecting income and consumption when he introduced the idea of the 'economic multiplier'; this is an idea which plays an important role in Keynes's *General theory.* The model was modified to take into account lags in the 'multiplier' by Kalecki (1935) and Hicks (1950). With the work of Goodwin, Tustin and Phillips, who all used diagrams to represent the structure of economic systems, the concept of feedback was made explicit.[4]

The concept of feedback is also implicit in the theory of natural selection. A. R. Wallace in an essay which he sent to Darwin wrote,

> The action of this principle [the struggle for existence] is exactly like that of the steam engine [governor], which checks and corrects any irregularities almost before they become evident; and in like manner no unbalanced deficiency in the animal kingdom can ever reach any conspicuous magnitude, because it would make itself felt at the very first step, by rendering existence difficult and extinction almost sure to follow.[5]

A simple example of this process – the balance between the populations of two kinds of fish in a closed pond – was given mathematical form by Volterra in 1931.[6]

At the same time that the theory of natural selection was being developed, a Frenchman, Claude Bernard (1813–1878), recognised that an organism reacts so as to keep both its internal and external environment constant, 'La fixité du milieu intérieur est la condition de la vie libre'.[7] He described several physiological mechanisms which exhibited feedback, but his ideas were ahead of his time and it was not until the 1920s, with the work of W. B. Cannon, that control mechanisms in the body began to be studied seriously. Cannon coined the name 'homeostasis' to describe these regulatory functions.[8]

Despite these instances, the way forward towards a clear understanding and a mathematical formulation of the theory of feedback systems was through engineering: first through mechanics — the regulation of prime movers led to an understanding of stability, the positioning of heavy loads to the development of servomechanisms — and then, as difficulties of analysis of mechanical systems began to hinder further progress, through electronics, and, in particular, through the need to obtain low distortion in the amplification and transmission of telephone signals.

This phase occupied 150 years, a period extending roughly from 1790 to 1940; progress was fitful, practical engineers often being far ahead of the theoretical understanding of what they were trying to achieve. It was to a large extent the period of the inventor. It was also a period when the subject boundaries were firmly held; a common control systems language had not been developed. The development of a coherent subject of control systems, beginning as it did in the 1930s, falls outside the scope of this book. It was during the World War II, with the need for servomechanisms to operate at higher speeds and with much greater precision than previously thought possible, that engineers and mathematicians came together to create the control engineer.

The concept of feedback, originally developed in political economy, has re-entered political language in the 20th century as part of a conscious attempt to emulate scientific method. And because man has always attempted to explain himself and his societies by metaphors drawn from the physical world, the metaphor has constantly changed to reflect the changing, progressively more complex, systems of the physical world. Thus, for example, God has been the 'great clock-maker in the sky', man his creation being but a complex mechanism; however, now it is 'being conjectured in biological circles that the laws of biology are to be conceived as computer programmes rather than as the systems of mathematical equations beloved of the physicist.'[9]

To Plato the flux of political life, the dynamic nature of society, was symptomatic of a diseased society. By organising political society on the basis of mathematics — number — stability could be obtained:

> The legislator must take it as a general principle that there is a universal usefulness in the subdivisions and complications of

number . . . All must be kept in view by the legislator in his injunction to all citizens, never, so far as they can help it, to rest short of this numerical standardization. For alike in domestic and public life and in all the arts and crafts there is no other single branch of education which has the same potent efficacy as the theory of numbers . . .[10]

Two thousand years later, Hobbes appealed to geometry, rather than to number, in an attempt to transform ethics and politics into exact sciences. And Leibnitz, like Hobbes fascinated by the geometrical method, tried by geometrical arguments to persuade Louis XIV to divert his armies from the Low Countries to Egypt.[11] But the reliance on geometry meant that the assumption still held that society was static.

Plato's pupil, Aristotle, was more concerned with the actualities of a political society than his master had been. To Aristotle the principle of a stable state was balance: the constitution should provide a balance between the forces of aristocracy and democracy.* This idea of balance found its metaphor in the machines of the eighteenth century: in explaining the principle underlying the American constitution, Alexander Hamilton exhorted, 'Make the system *complete* in its structure, give a *perfect proportion* and *balance* to its *parts,* and the power you give it will never affect your security.'[12] But in view of the dynamic nature of political society, 'Because politics resembles ballet more than it does a still-life painting, any description must emphasise movement and the relationship among the parts of the ensemble',[13] a view which may have been implicit in earlier political theorists, but becoming explicit at least from Machiavelli onwards,

---

* The implications of Aristotle's view are clearly expressed in comments on Machiavelli, analysing politics in the Aristotelian tradition: 'To Aristotle, the *polis,* or the best possible state, consisted in the mixture of two elements, not the philosophical or the practical dominance of a single element: aristocracy, seen in principle as the rule of the wise, constitutionally as the rule of a few and socially as the rule of the wealthy or those with property; and democracy, seen in principle as the rule of mere opinion . . . constitutionally as the rule of the many, and socially as the rule of the poor. Neither principle alone is likely to render either stable or just government: they need to be synthesised, and this synthesis is not a mere compromise, but the taking of the two best elements in both, wisdom and consent, and making them into a new, organic unity . . . . The mixture, however, in both Aristotle and Cicero is somewhat static . . . Machiavelli first introduces a dynamic element into a theory of mixed government . . . . So the mixture is not a static matter of checks and balances, but a dynamic blending of three elements that are needed in different proportions at different times . . .'

Crick, B. R. C.: 'Introduction' in Machiavelli, N: *The Discourses* (Penguin, Harmondsworth, 1970), pp. 23, 29.

and we should not be surprised that political scientists now turn to control systems and communications systems for their models,[14] or that politicians use the language of control systems:

> The industrial mentality thinks that efficiency comes from centralisation of power, ignoring the ever-increasing need for information in the system, and especially the need for negative feedback so that poor plans can continually be corrected before errors mushroom into disasters.[15]

For the social sciences, the 20th-century metaphor is the 'feedback system'.

References and notes

1 HALL, H. R.: *Governors and governing mechanism* (The Technical Publishing Co., Manchester, 2nd edn. 1907), p. 8

2 DAVID HUME: 'Of the balance of trade', 1752, quoted from MAYR, O.: 'Adam Smith and the concept of feedback', *Technology & Culture*, 1971, 12, pp. 3–4

3 *ibid.*, p. 6

4 STONE, R.: *Mathematics in the social sciences and other essays* (Chapman & Hall, London, 1966), pp. 34–37; KAHN, R. F.: 'The relation of home investment to unemployment', *The Economist Journal*, 1931, XLI, pp. 173–98; KEYNES, J. M.: *The general theory of employment, interest and money* (Macmillan, London, 1936); KALECKI, M.: 'A macrodynamic theory of business cycles', *Econometrica*, 1935, III, pp. 327–44; HICKS, J. R.: *A contribution to the theory of the trade cycle* (Oxford University Press, 1950); GOODWIN, R.: 'The nonlinear accelerator and the persistance of business cycles', *Econometrica*, 1951, 19, pp. 1–17; TUSTIN, A.: *The mechanism of economic systems* (Heinemann, London, 1953); PHILLIPS, A. W.: 'Stabilization policy and time-forms of lagged responses', *The Economic Journal*, 1957, LXVII, pp. 265–77; PHILLIPS, A. W.: 'Stabilization policy in a closed economy', *The Economic Journal*, 1954, LXIV, pp. 290–323

5 Quoted from BATESON, G.: 'Conscious purpose versus nature' in *The dialectics of liberation*, Cooper, D. (ed.), (Penguin Books, Harmondsworth, 1968), pp. 36–37

6 VOLTERRA, V.: *Théorie mathématique de la lutte pour la vie* (Gauthier-Villars, Paris, 1931); for a brief account, see MINORSKY, N.: *Non-linear oscillations* (Van Nostrand, Princeton, 1962), p. 65

7 Quoted from GREY WALTER, W.: *The living brain* (Penguin Books, Harmondsworth, 1961), p. 41; Bernard's book *Introduction to the study of experimental medicine* was translated into English by H. C. Green and published in New York in 1949

8 A summary of his work is in CANNON, W. B.: *The wisdom of the body* (W. W. Norton, New York, 1932)

9 FLOWERS, B. H.: 'Technology and man; the first Leverhulme memorial lecture, 25th October, 1971 (Liverpool University Press, 1972), p.10

10 *The laws of Plato*, translated by TAYLOR, A. E. (Dent, London, 1934), p. 747, quoted from WOLIN, S. S.: *Politics and vision* (Little, Brown, Boston, 1960), p. 49

11 COHEN, J.: *Human robots in myth and science* (George Allen & Unwin, London, 1966), pp. 76–77

12 HAMILTON, A., quoted from MORRIS, R. B.: *Alexander Hamilton and the founding of the nation* (Harper & Row, New York, 1969), p. 238

13 ROSE, R.: *Politics in England today* (Little, Brown, Boston, 2nd edn., 1974), p. 16

14 See, for example, EASTON, D.: *A systems analysis of political life* (Wiley, New York, 1965); YOUNG, O. R.: *Systems of political science* (Prentice-Hall, Englewood Cliffs, 1968); DEUTSCH, K. W.: *Nerves of government* (Free Press, New York, 1967)

15 ERIC MOONMAN, *The Times*, 18th June 1975

Chapter 2

The regulation of
prime movers

The governor is of a nature solely calculated to secure more
effectually an equable motion under different degrees of heat
from the fire; a property so extremely essential in preparing
cotton to work into fine yarn . . .

Peter Drinkwater, in a letter to Boulton and Watt, 1789

Introduction

Common usage in the 19th century applied the term prime mover to
the windmill, water wheel and the steam engine, notably to the latter.[1]
It is here extended to include water and steam turbines, clockwork
devices and weight-driven apparatus.

With the decline of the Roman Empire, Mumford's 'megamachine'[2]
began to be replaced by animal, wind and water machines, but man
remained a part of the control loop. It was only with the development
of mechanical clocks during the Middle Ages that systematic efforts to
develop speed regulation began.[3] The early mechanical clocks were
weight driven and regulated by a rotating vane attached to the clock
drive; the resistance offered by the vane, owing to the frictional resistance of the air, depended on the speed of rotation, and hence this
resistance provided a feedback path. Gradually, the rotating-vane
regulator was replaced by the verge and foliot mechanism which prevented the weight from accelerating as it fell by stopping the motion
at frequent intervals. Like the float valve, the rotating vane is a machine
part which was to re-appear in a variety of forms (at the beginning of
the 19th century it was known as the Venetian fly) and can still
occasionally be found in use as a speed regulator.

In the latter part of the 17th century Christiaan Huygens and Robert

Hooke made extensive investigations into the problems of accurate timekeeping. Huygens showed how a simple pendulum could be modified to have a period of oscillation independent of the amplitude of swing and he also determined the surface of revolution necessary for the period of a conical pendulum to be independent of its radius of rotation. Both men designed clocks based on the use of the isochronous conical pendulum as a regulator, and Hooke also designed a weight-driven equatorial telescope that was to be regulated by an ordinary conical pendulum (Fig. 2.1). Fuller has suggested that the investigations concerning the isochronous conical pendulum represent the beginnings of control theory, the isochronous pendulum being a controller without offset.[4] With the development at the end of the 17th century of the

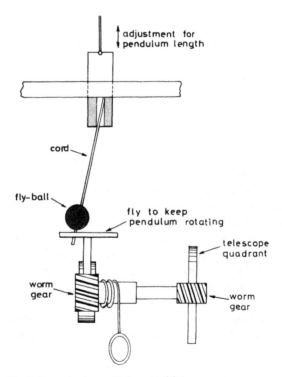

Fig. 2.1   *Hooke's constant-speed telescope drive*

anchor escapement, which limited the amplitude of oscillation of the clock pendulum, there was no longer a need for a truly isochronous pendulum and the attention of the clockmaker turned towards the problem of obtaining a pendulum whose length was independent of variations in temperature.

In 1712 the two major sources of power, wind and water, were complemented by a third — steam. Newcomen's steam engine had been developed to meet the needs of mine owners, and, for the next 70 years, the majority of the steam engines built by Smeaton, Watt and others were pumping engines for drainage. The major form of control was the cataract which regulated the stroke frequency; other controls were weight-loaded safety valves used on the boilers, float valves used to regulate the boiler feed water and eventually boiler pressure regulators.[5]

In the 1770s the range of applications widened and energy-storage devices — pneumatic acculumators and flywheels — came into use. John Smeaton even advocated a pump storage scheme, whereby a steam engine, instead of providing rotary motion directly, was used to pump water to a reservoir, the rotary motion then being provided by a water wheel. An advantage of this scheme was that by the nature of its construction a water wheel, like a windmill, possessed high inertia: an inbuilt flywheel. The regulation properties of the flywheel had long been known: the heavy potters' wheel and the disc attached to a bow drill date from palaeolithic times; the flywheel was recorded as a separate item of machinery in a 15th-century manuscript, and there are frequent references to the use of 'flys' in technical literature.[6] It is surprising, therefore, that a number of attempts were made to produce rotative steam engines without flywheels. Watt in later life recalled that in the late 1770s he had been 'desirous to render the motion equable without a flywheel, the regulating powers of which I did not then fully appreciate.'[7] By 1782, however, Watt was concluding that 'the flyer is the best of all and will prove the true equaliser'.[8]

Despite the growing importance of the steam engine, the major sources of power during the 18th century were the water wheel and the windmill. Of these, the latter was the most unpredictable in its operation, being dependent on both the direction and strength of the wind. It is not surprising that several feedback mechanisms designed to secure uniform motion, or to compensate for the irregularity of the motion of the windmill, were developed during this period. The earliest of these feedback mechanisms, the fan tail and the shutter sail, were described in a patent granted to Edmund Lee in 1745.[9] The fan tail, which was used to keep the windmill sails facing into the wind, became widely used during the 19th century. The shutter-sail mechanism was designed to be a speed regulator; the cloth sails were replaced by shutters, pivoted at each end and held shut by a counterweight. If the wind force increased, the shutters would be blown open and hence the torque generated would be reduced. This was not a true speed regulator: the controlled variable was output torque not speed, but, in a modified form, with spring-loaded shutters, it provided an adequate method of speed control for several generations of millers.[10]

The miller's major concern was not the achievement of some specified speed, but the maintenance of the correct gap between the upper and lower grindstones. An increase in speed causes the top stone to rise, the increased gap giving coarser flour; in gusty conditions the miller had to be continually testing the flow and adjusting the gap, 'tentering' as it was called. In 1785[11] Robert Hilton suggested a method of automatically adjusting the gap according to the speed of the mill. A centrifugal fan driven by the mill with a baffle placed in the output from the fan was to operate the 'lift tenter', the name by which the tentering gear was known. Hilton's lift tenter does not seem to have been used in practice; however, the idea of using a centrifugal fan as a speed-sensing device was revived in the 19th century by Foucault and by Charles Parsons.

The practical lift tenter was the invention of Thomas Mead, who was granted a patent for it in 1787. Mead's lift tenter was based on the use of a double conical pendulum; the height of the fly balls depended on the mill speed, and, by means of levers, variations in this height were transmitted to the lift tenter. The lift tenter was not a speed regulator: it was part of an open-loop system for adjusting the gap between the stones in accordance with variations in speed.

Mead, in his patent, also proposed a closed-loop speed regulator for 'the better and more regular Furling and Unfurling Sails on Wind-mills without the constant Attendance of a Man.'[12] This regulator is shown in Fig. 2.2. The change in height of the conical pendulum moves a sliding collar on the main shaft, and ropes connect this collar with the sail furling mechanism. At zero speed the sails are held fully unfurled by helical springs acting on the collar. As the speed increases so does the required centrifugal force; this force is provided by the extension of the helical spring, and as the spring extends the sails are furled. The only evidence that Mead's speed regulator was ever used comes from a Dr. Alderson, who, in an address given at the Hull Mechanics' Institute in 1825 said, 'The principle [of the centrifugal governor] was borrowed from the patents of my late friend Mead, who, long before Mr. Watt had adapted the plan to the steam engine, had regulated the mill-sails in this neighbourhood upon that precise principle, and which continued to be so regulated to this day.'[13]

The lift tenter, however, was extensively used; it can still be seen

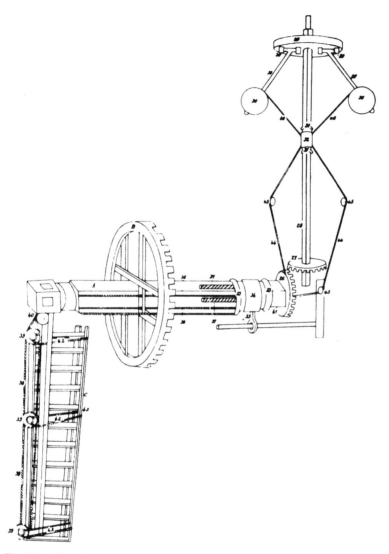

Fig. 2.2   *Mead's speed regulator for windmills*
[Reprinted from *The origins of feedback control* by O. Mayr by permission of the MIT Press, Cambridge, Massachusetts. © The MIT Press, 1970]

in the few remaining operational windmills.[14] Evidence collected by Watt[15] in preparation for possible patent litigation suggests that the lift tenter operated by a centrifugal pendulum was in widespread use well before the time of Mead's patent. There is, however, no mention of it in Smeaton's windmill design of 1782,[16] and it is not until 1788, after the issue of the patent, that Boulton describes its use at Albion Mills.

During the early 1780s, Boulton and Watt were active in promoting the commercial use of the rotative steam engine, and it was partly for publicity purposes that they decided to construct and operate the Albion flour mills, located close to Blackfriars Bridge in London.* They obtained their publicity, although Watt was none too happy about it: 'It has given me the utmost pain to hear of the many persons who have been admitted to the Albion Mill merely as an Object of Curiosity', he wrote in a letter to Boulton in 1786.[17] However, initial enthusiasm waned and by the time the mills were destroyed by fire in 1791 the technical innovation was being blamed for the increase in the price of flour, which had, in fact, been brought about by a rise in the price of wheat.

As a commercial venture the Albion Mills were a failure. Boulton and Watt lost between £6000 and £9000, but out of the experience came many new ideas. In 1788 technical problems with one of the engines drew Matthew Boulton to London, and, while there, he saw the improvements to the milling machinery which had been made by John Rennie. In a letter to Watt dated 28 May 1788[18] he described some of the new machinery, including a mechanism

> for regulating the pressure or distance of the top mill stone from the Bed stone in such a manner that the faster the engine goes the lower or closer it grinds & when the engine stops the top stone rises up & I think the principal advantage of this invention is in making it easy to set the engine to work because the top stone cannot press upon the lower until the mill is in full motion; this is produced by the centrifugal force of 2 lead weights which rise up horizontal when in motion & fall down when ye motion is decreased, by which means they act on a lever that is divided as 30 to 1, but to explain it requires a drawing.

---

* The firm of Boulton and Watt began in 1783 to build, in London, a steam-powered flour mill, known as Albion Mills. The mill began operating with one engine in 1786, and the second engine was not brought into operation until 1789. Boulton made frequent visits and it is therefore unlikely that the lift tenter was in operation much before 1788; the engineer employed to supervise the installation of the engines and the mill work was John Rennie and it is also unlikely that he was unaware of the latest developments in mill machinery, all of which strongly suggests that the lift tenter was not in widespread use prior to 1787. For details of Boulton and Watt's involvement in the Albion Mills, see Dickinson and Jenkins,[15] pp. 64, 65, 123, 167, 220–2, and Roll, E.: *An early experiment in industrial organisation: being a history of the firm of Boulton & Watt, 1775–1805* (Longmans, Green, London, 1930), pp. 111–4.

The timing of Boulton's letter was excellent; Watt had just perfected a simple and convenient means of manual regulation of the engine, namely, the throttle valve. This valve was light, requiring little force to operate, and Watt quickly recognised that in combination with the lift tenter described by Boulton it would provide a mechanism for automatically setting the engine speed. Work soon began on designing the so called fly-ball or centrifugal governor; a drawing with the title 'Centrifugal speed regulator' is dated November 1788, and, in December 1788, a drawing of the governor for the famous 'LAP' engine was produced (Fig. 2.3).

The arrival of this practical engine governor was unheralded. There was no announcement in scientific journals and there was no attempt to patent it; Boulton and Watt adopted a policy of secrecy. The first description of a centrifugal governor appears to be that published in Nicholson's Journal in 1798; this referred to a governor applied to the regulation of a water wheel.[19] In 1807 Thomas Young described the governor and its application.[20] Nevertheless, the news spread. A customer, Peter Drinkwater, wrote to Boulton and Watt in 1789, to say that the governor 'is of a nature solely calculated to secure more effectually an equable motion under different degrees of heat from the fire; a property so extremely essential in preparing cotton to work into fine yarn, that I would on no account have you deny the use of this instrument.'[21] In 1790 John Rennie, in London, was asking for four or five units, and in 1793 a competitor copied the governor.

Governors with integral action

The widening use of the Watt governor, particularly in its application to the regulation of water wheels, revealed two defects in its action: hunting and lack of power. Preuss in 1823 suggested that hunting was inherent in the action of the conical pendulum which he considered could only be in equilibrium at the design speed; an increase or decrease in speed, however small, would cause the governor to move to its maximum or minimum position: 'Hence the speed must oscillate between the maximum and the minimum; while, in order to have an equal motion of the machine, the difference between the two possible extremes ought to be as small as can be'.[22] Preuss had a mistaken view of the action of the governor; he considered only the action of the centrifugal force and ignored the changes in the restoring force which accompany a change in the radius of the circle of rotation. He

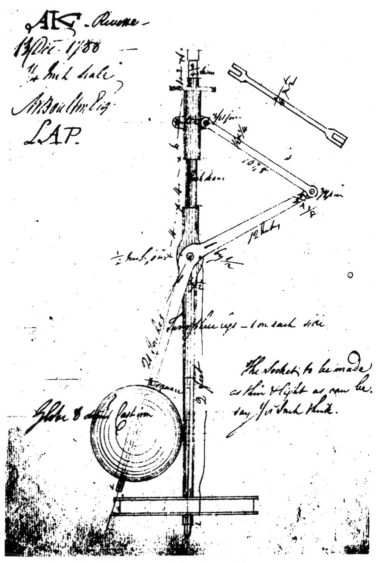

Fig. 2.3   *Drawing of governor for LAP engine, 1788*
[Courtesy of the City Librarian, from the Boulton and Watt Collection, Birmingham Public Libraries]

'endeavoured to invent a contrivance which might not be subjected to the inconveniences' of the centrifugal governor. His invention was, in fact, a variation on the pump regulator of the Perier brothers.

Hannuic, to whom the British Patent rights of a pneumatic regulator were granted, held similar views. He wrote of the centrifugal governor that

> a constant vibration of the balls is produced and only a slight approximation to regularity of speed obtained . . . . The insufficiency of the centrifugal regulator, as applied to water wheels, is too well known to need any comment, the shuttle or water gate being in most cases too ponderous to be affected by any force gained by vibration of the balls.[23]

Pump regulators

To avoid the defects of the Watt governor, various forms of hydraulic and pneumatic pump regulators were produced. It is thought that the pump regulator was first developed by Watt as an alternative to the cataract used on pumping engines.[24] A hydraulic form of the pump regulator was described by de Prony and attributed by him to the Perier brothers; its action is described in detail by Mayr.[25]

The principle of the hydraulic or pneumatic regulator is illustrated in Fig. 2.4. The use of the accumulator provides *integral* action, and hence one defect of the Watt governor — offset — is avoided. By the use of loaded accumulators, the high forces needed to operate sluice gates on water wheels could be obtained; the evidence suggests that it was the latter advantage, not the elimination of offset, which led to the extensive use, particularly in France, of the pump regulator.[26]

Louis Molinié, a cotton manufacturer from Saint-Pons, France, patented one such regulator in 1837[27] (the British Patent rights to Molinié's regulator were granted to Hannuic). A pump driven by the engine supplied air to an inflatable reservoir, the pressure in which was determined by means of weights attached to the top. A small orifice in the reservoir allowed the air to escape at an almost constant rate, since the pressure was kept constant. If the speed of the engine deviated from the nominal the reservoir was inflated or deflated and this motion was transmitted to the valve controlling the engine.

In principle, the controller acted as an integrator, but the elasticity of the reservoir resulted in a transfer function of the form

$$\frac{k}{c_0 + s} \tag{2.1}$$

The controller also incorporated a positive feedback loop, in that the size of the orifice was made to depend on the height of the weight. This had the effect of reducing c_0 in eqn. 2.1 to zero, thus giving true integral action. An investigation of the action of Molinié's regulator carried out by Combes in 1841 showed that it had a settling time of several minutes in some cases, indicating that the system was close to instability.[28] This was probably the first measurement ever made of the transient response of a feedback system.

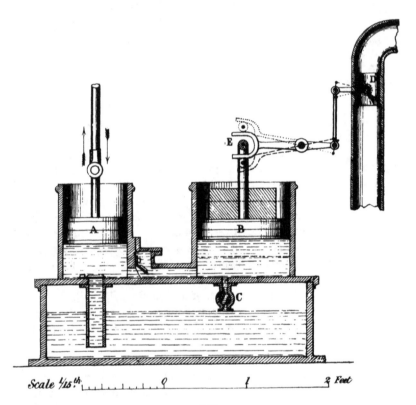

Fig. 2.4   *Hydraulic pump regulator, 1853*
[Reprinted by permission of the Council of the Institution of Mechanical Engineers from *Proc. I Mech. E*]

The practical significance of Molinié's regulator, however, was not its integral action, but the fact that it could generate much larger forces than the centrifugal governor, and hence could be used to regulate water wheels. In 1841 it was reported that over 400 water wheels had been fitted with his apparatus.[29] A similar device referred to as Heinrich's governor was mentioned in the discussion of a paper by Woods in 1846.[30]

Speed-reference governors

A governor on a very different principle was patented by Benjamin Hick in 1840.[31] He attempted to devise regulators in which the actual and desired speeds of the engine were compared directly, rather than in terms of forces, as with the centrifugal governor. To do this he required a source of constant speed to provide the reference. Of the three methods which he proposes in his patent application, the method shown in Fig. 2.5 is perhaps the most ingenious. The heavy nut, which is loosely fitted on the vertical screw, descends under the action of gravity and, in doing so, rotates. The rate of rotation is checked by the air resistance of two large vanes attached to the nut which consequently rotates at a roughly constant rate. The vertical axle is driven by the engine, and if the speed of the engine matches the rate of rotation of the nut, the nut remains stationary. If the engine speed deviates, however, the nut will move up or down the vertical screw; this movement can be used to operate the throttle valve.

Although Hick makes no mention of the fact, his governor incorporates integral action. A difference in speed between the axle and the nut causes the nut to move along the axle with a *velocity* proportional to the difference. The vertical *displacement* is therefore the integral of the error in engine speed. Because the vanes are loosely coupled to the engine shaft they will tend to maintain the momentary engine speed. If, owing to sliding-friction effects in the screw, this momentary running speed is greater than the free-descent speed of the vanes, some differential action will be introduced. In practice, therefore, a Hicks governor would have had integral action plus a small amount of differential action.

In 1805 Rider[32] had patented a method involving direct speed comparison using as a reference source the most accurate constant speed device available, the pendulum. He arranged for the engine to raise a weight, as if it were winding up a clock, and an oscillating pendulum and escapement mechanism were used to regulate the lowering of the weight; hence, when the engine was running at the set speed, the weight was stationary, but any variation in speed caused the weight to rise or fall, and this motion was transmitted to the steam valve. It is not known whether this mechanism was ever used in practice; it is referred to by Rees[33] (1819) and Field[34] (1846) and is a forerunner of the chronometric governor of the Siemens brothers.

Siemens's chronometric governor[35]

C. William Siemens (1823–1883) arrived in London in 1844 from Germany, with the chief purpose of selling the chronometric governor

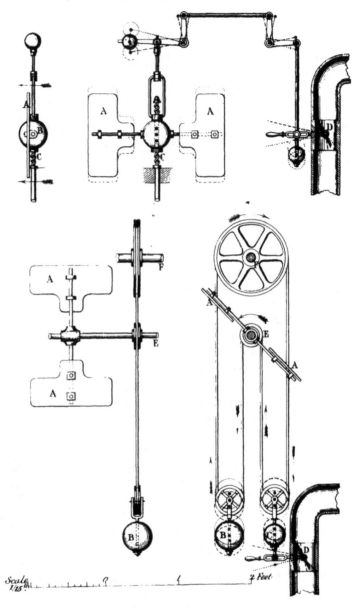

Fig. 2.5   *Hick's engine governors, 1853*
Upper: Hick's fly governor
Lower: Hick's fly governor, second form
[Reprinted by permission of the Council of the Institution of Mechanical Engineers from *Proc. I Mech. E*]

invented jointly with his brother E. Werner Siemens (1816–1892). To assist in publicising the invention, William Siemens obtained the help of Joseph Woods, a civil engineer, in whose name the British Patent was granted,[36] and who in 1846 presented a paper on the governor to a meeting of the Institution of Civil Engineers.[37]

The Siemens brothers recognised two defects in the Watt and other types of centrifugal governor. The first was that 'it does not regulate, but only moderates the velocity of the engine'[38] and the second was described as a defect common to both centrifugal and other forms of governor: 'insensibility, or want of power to act, at the moment when a change takes place in the power or the load'.[39] The effect of this want of power on the operation of the Watt governor was described by Woods:

> But having to pull the valve, and to overcome the friction in all the joints, the weighted levers will not alter their arc of vibration with sufficient rapidity. They will vary their velocity in concert with the engine, without acting upon the valve, until after the engine has altered its speed sufficiently to impart so much centrifugal power to the balls, as will cause them to overcome the resistances. The flywheel of the engine will, in the mean time, have altered its momentum to such an extent that the different supply of steam will not immediately counterbalance it; the governor balls will, consequently, alter their arc of vibration more than is really required, and the speed of the engine will be caused to fluctuate for a short time.[40]

Here, clearly, is the recognition that frictional forces in the throttle valve and linkages could give rise to 'hunting'.

In contrast, the chronometric governor responded quickly to changes in engine speed, as Werner Siemens explained:

> With the centrifugal governor an acceleration of the speed of the engine by one-twentieth of a revolution cannot be indicated, because the centrifugal force of the balls is not sufficiently increased by this slight increase in the velocity of revolution to overcome the resistance opposed to their flying out. The differential governor has completed its full action, and again completely regulated the speed of the engine, before the centrifugal governor has ever begun to move.[41]

This quickness of response was a consequence of using the power of the engine to move the throttle valve: the governor was acting as a mechanical amplifier.

The Siemens governor seems to have been developed from the notion that if the conical pendulum is freed from the engine it will rotate in smaller and more isochronous revolutions, and that if a sudden change to the engine load is made such that the engine

> begins to move more quickly or more slowly, the free swinging pendulum, which retains its original velocity, must either lag behind or gain. . . . the regulation of the speed of the engine is made to depend on the diversity of the travel of the engine and governor in equal times, or rather on the difference of the two.[42]

The brothers then apparently proceeded with a systematic examination of ways of obtaining the difference in travel of the engine and pendulum, eventually adopting the differential gear mechanism shown in Fig. 2.6.

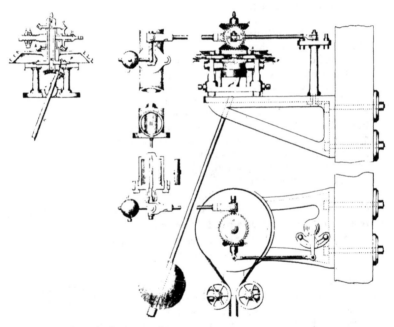

Fig. 2.6   *Siemen's chronometric governor*
[*Minutes and Proceedings of the Institution of Civil Engineers*, 1846, 5, Pl. 19. Courtesy of the Institution of Civil Engineers]

If the upper bevel gear is held stationary the weight will drive round the pendulum, and hence the system will behave like a friction governor; in normal operation, however, the weight remains fixed and the upper bevel wheel rotates about its own axis, and this in turn drives the pendulum. The power to overcome the friction in the pendulum thus comes from the engine and not from the weight. As the engine speeds up, the pendulum bob is forced against the fixed brake surface, and the reaction torque thus generated is applied via the intermediate bevel wheel to the control linkage. If the engine speed drops below the nominal value, the inertia of the pendulum will initially tend to keep it rotating at the nominal speed, and the weight will move the intermediate wheel about the vertical axis and hence operate the control linkage. Should this condition persist, the weight also has to provide the driving torque to maintain the pendulum in motion.

The brothers also realised that their governor, as well as being quicker in action than the common governor, differed in a more significant way, since 'a centrifugal governor can only alter but cannot completely stop the existing difference of motion of the engine, [whereas] the differential governor absolutely compels it to take up again the speed determined on.'[43] That is, the chronometric governor eliminated offset. The elimination of offset was a direct consequence of the use

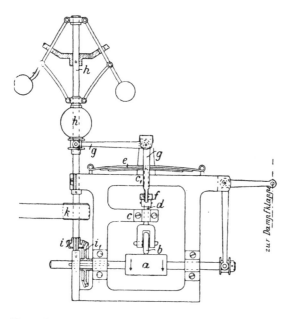

Fig. 2.7 *Werner Siemen's integral action governor*
[Scientific and technical papers of Werner von Siemens]

of a differential gear to compare the reference speed with the engine speed. The gear mechanism automatically introduces integral action since the comparison is made in terms of distance travelled, not speed.

William Siemens in all his writings on governors continued to emphasise this distinction [Siemens's italics]: 'It [the Watt governor] cannot *regulate*, but only *moderates* the velocity of the engine that is, it cannot prevent a permanent change in the velocity of the engine when a permanent change is made in the load upon the engine'.[44]

Over thirty years later, in 1882, Werner Siemens designed a governor in which integral action was introduced by the deliberate manipulation of the error signal obtained from an ordinary proportional-action governor.[45] The apparatus is shown in Fig. 2.7. The movement of the governor collar, which is proportional to the difference between the engine speed and the set speed, is integrated by a wheel-and-cylinder mechanical integrator.[46] The wheel is fixed and the cylinder is free to move axially; the forces between the wheel and cylinder cause the cylinder to move axially at a rate proportional to the angle of inclination of the wheel.

In England, William Siemens came to exercise considerable influence among scientists and the more scientific engineers on questions of engine governing. The chronometric governor had been received with great acclaim: the Astronomer Royal, Sir George Biddell Airy, adopted it for use on equatorial telescopes, and, through Airy, Siemens was drawn into the scientific circle that included Maxwell, William Thomson and Edward John Routh.[47]

Commercially, however, the chronometric governor was a failure. An improved version was demonstrated at a meeting of the Institution of Mechanical Engineers in 1853 (Fig. 2.8).[48] This new governor was more powerful than the original governor – the model demonstrated could support a weight of 1·5 cwt (76 kg) on the throttle lever – it was also suitable for use at sea, since it could operate in a slanting position. The improvements did not, however, obviate the inherent weaknesses in this type of governor, namely wear of the friction brake and high-frequency oscillation. The oscillatory motion, although it did not disturb the steam engine, tended to deter potential customers. It also aggravated the problem of wear. Siemens eventually abandoned the mechanical-friction governor, and in 1866 designed a new governor based on a hydraulic brake.[49]

The dream of perfection

> ... during an experience of 25 years I have always found the old-fashioned pendulum governor, as used by Watt, to be quite sufficient to regulate the motion of the steam engine, yet from

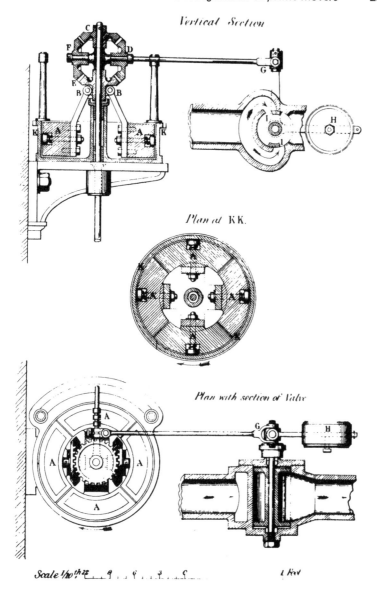

Vertical Section

Plan at KK.

Plan with section of Valve

Scale 1/10th ...

Fig. 2.8   *Siemen's improved chronometric governor, 1853*
[Reprinted by permission of the Council of the Institution of Mechanical Engineers from *Proc. I Mech. E*]

want of attention to the mode of adjustment the presence of
75% of these governors is more ornamental than useful. Hence
the great influx of patent and other governors, some of which
are neither ornamental or useful . . .

T. Patterson, in a letter to *The Engineer*, 1868, **26**, p. 24

Several systems [electricity supply companies], which have
investigated governor performance at various times, have been
convinced that many of the troubles can be laid to inadequate
maintenance of the governors.

J. J. Dougherty *et al.*, *Trans. AIEE*, 1941, p. 554

Isochronous governors*
It has been estimated that, in 1868, the year in which Patterson's letter
was written, over 75 000 governors of the Watt type were in use in
England.[50] They were used on the steam engine, waterwheels, telescope
drives and were soon to be used on steam and water turbines, telegraphic
apparatus, the phonograph, oil and gas engines and for the speed control of electric motors. With this vast market, many inventors dreamed
of fortunes: over a thousand patents for governors (excluding shaft
governors) were granted in the USA between 1836 and 1900, although
few, of course, ever went beyond the patent-model stage and fewer still
were commercially successful.[51] For, although the Watt governor was
'imperfect and partial in its action', it was 'sufficient for many purposes;
and its extreme simplicity has caused it to remain to the present time
in more extensive use than any other.'[52]

The eminent French physicist J. B. L. Foucault[53] was one such
inventor. In 1862 he produced a friction governor based on the aerodynamic resistance of a centrifugal fan,[54] a device reminiscent of
Robert Hilton's lift tenter. This governor apparently worked well and
Foucault had visions 'of triumphing where James Watt had only a
partial victory', visions of great commercial success and financial
independence. But his vision was not to be realised; in industrial terms
Foucault's governors remained oddities, and he never achieved the
financial independence for which he yearned.[55]

----

* The term 'isochronism' (*isochronisme*), meaning the absence of offset in a
speed-control system, gained wide currency, in French literature, from its first
use by Pecqueur and Foucault in 1847. See Foucault, L.: 'Sur une horloge à
pendule conique', *Comptes Rendus*, 1847, **25**, p. 159 and Pecqueur, 'Sur un
Pendule Centrifuge à Isochronisme Naturel', *Comptes Rendus*, 1847, **25**,
pp. 251–253. In German literature the term 'astatic' rather than 'isochronous'
was used, it was introduced by F. Realeaux in 1859, 'Zur Regulatorfrage',
Zeitschrift des Vereines deutscher Ingenieure, 1855, **3**, pp. 165–168.

Like many other inventors around the middle of the century, Foucault was searching for an 'isochronous' governor, a governor without offset. Charbonnier in 1843 had devised a loaded governor in which the load was applied by a bent lever in such a manner that the restoring force and centrifugal force were balanced at all positions.[56] Similar governors were designed by Marie Joseph Farcot in 1854 and by Foucault in 1862.[57] In the 1850s, governors in which isochronism was obtained by constraining the bob of the conical pendulum so that it described a paraboloid of revolution were devised. One such governor was demonstrated at the Great Exhibition in 1851, and Farcot designed a similar governor in 1854. Together with his son, Jean Joseph Léon Farcot, he designed several more governors.* Among these was a crossed-arm governor (Fig. 2.9), which, by a suitable disposition of the

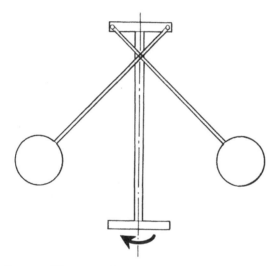

Fig. 2.9   *Farcot and Son, crossed-arm governor*

pivot points can be made approximately isochronous. The Farcots found that, in operation, their governor sometimes went 'mad', i.e. unstable. To control the oscillation they tried first a simple friction brake, and, in 1864, an air dashpot.

Other engineers had similar stability problems with isochronous governors. In 1871 an English engineer, J. Head, wrote of one such

---

* Jean Joseph Léon Farcot is remembered not for his work on governors, but as the author of the first book on servomechanisms, published in 1873 (see chapter 4).

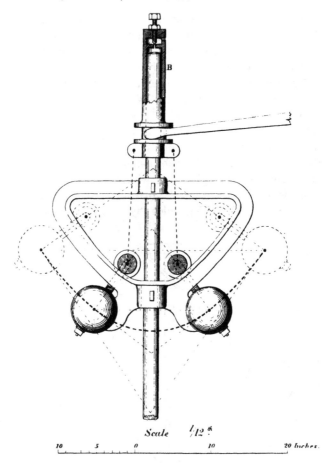

Fig. 2.10   Parabolic governor with air damper, 1871
[Reprinted by permission of the Council of the Institution of Mechanical Engineers from Proc. I Mech. E]

governor (Fig. 2.10) that

> had there been no momentum in the moving parts . . . the action
> of the governor would have been perfect. But it was rendered
> completely useless by the continued disturbance of the throttle
> valve, after the proper position had been reached . . . . This dif-
> ficulty was then obviated by a very simple expedient, namely
> by attaching to the sliding-piece of the governor an air cylinder . . .
> In rising or falling, the balls were made to suck or force air into
> or out of a small adjustable aperature in the top of this cylinder;
> and in this way no velocity or momentum in the vertical direction
> was allowed to accumulate. The parabolic governor so modified
> was perfectly successful, and after fifteen years' working is still
> in operation . . .[58]

Foucault also observed that the use of isochronous governors tended to make engine control systems unstable, but unlike some engineers whom, he said, believed that instability was an inevitable consequence of the neutral equilibrium of an isochronous regulator, he thought that instability could be avoided by reducing backlash in the gears, slackness in driving belts and similar faults in the transmission between the engine and the governor.[59] It seems to have been obligatory, in France at least, for writers on governors to discuss in detail Foucault's inventions, mainly because of his eminence as a physicist, rather than for any intrinsic merit they might possess. Rolland (1867, 1868, 1870),[60] Romilly (1872),[61] Villarceau (1872)[62] and Marie (1878)[63] all devote considerable attention to Foucault's work. By 1880, however, interest in isochronous governors was declining.

Dynamometric or load-sensing governors

A perfect governor must not be called into action by a change in speed, but must *feel* the cause of such a change, and anticipate its effect, making the necessary adjustment *before* the threatened alteration in speed actually takes place.

Mr. E. Hunt, in a letter to *The Engineer*, 1858, p. 169

As has been noted, not only did the common governor give rise to steady-state errors, but it was frequently slow in its response. An early attempt to obtain a faster response was that made by Jean-Victor Poncelet (1788–1867), a French army officer. Poncelet served under Napoleon, was captured during the retreat from Moscow and, while in captivity, made major contributions to projective geometry. In 1824 he was ordered to set up a course on the science of machines at the military school at Metz, and for this course produced a set of notes which were lithographed in 1826, 1832 and 1836 before receiving full publication in 1874.[64] In the 1836 edition, Poncelet proposed a new form of engine governor which was based on the use of *feedforward*. The controller was to operate by measuring the load on the engine. A spring coupling was placed between the engine and load, and the twist in the coupling was used to adjust the throttle valve.[65] But, as Trinks noted, the drawback to Poncelet's proposal was in the inability of the controller to compensate for variations in the energy supply to the prime mover and also in the fact that the use of a flexible coupling could lead to 'never-ending vibrations'.[66]

By the middle of the century 'A Marine Engine Governor to check the admission of steam when the engine runs away, from the water leaving the wheels or screw [was] much required .... The ordinary governor has been tried for the purpose, but was found not to answer, as its action is too slow in shutting-off and letting-on steam.'[67] Several remedies were proposed. Waddell devised a scheme whereby the common governor was assisted in closing the throttle valve by the use of a piston activated by the pressure drop across the throttle valve, while a Mr Peter Jensen[68] in 1859, and a Mr Dunlop in 1877, devised a governor which 'worked by means of a siphon and small air chamber in the stern of the vessel, the air of which was compressed or expanded according to the depths at which the stern was submerged or free, action being communicated to the throttle valve accordingly'.[69] Another type was that designed by Vasserat in the 1850s. This governor was an open-loop device that used an ordinary pendulum to sense the pitching of the ship, and hence to adjust the throttle valve.[70]

Practical marine-engine governors being developed during the period included the 4-ball governor of Thomas Silver (1813–1888), patented in 1855, in which the effects of gravity were cancelled by extending the arms of a conventional conical pendulum upwards and adding at the upper end weights equivalent to the fly balls, the restoring force being provided by a spring.[71] Other types included the various spring-loaded governors similar to that designed by Brunel in 1822 (Fig. 2.11).

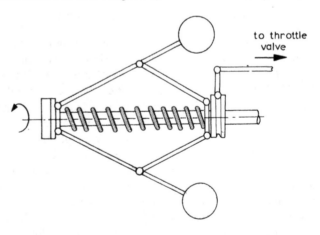

Fig. 2.11   *Brunel's spring-loaded marine-engine governor*

The improvement in the speed of response so necessary for marine use was not to come until towards the end of the century with the development of shaft governors incorporating proportional-plus-derivative action. A forerunner of these governors was that patented by H. N. Throop in 1857 specifically for marine-engine use. This device used several weights placed around the circumference of a frame that could move both tangentially and radially, the radial movement being proportional to speed and the tangential motion being proportional to acceleration, so that the governor provided proportional-plus-derivative action.[72]

Towards the end of the 19th century, designs for controllers incorporating a centrifugal governor plus a control signal obtained from a torque meter began to appear. Of one such controller, produced in 1883, Ball commented that 'remarkable results were thus obtained which, in some respects, have never been surpassed.'[73] The use of a similar controller for the speed control of electric motors was suggested by S. P. Thompson, who also noted that such a controller could be set to have a characteristic which would cause the motor to run faster with increasing load.[74] In recent years there has been a revival of interest in the use of governors with load sensing.[75]

Development of the practical engine governor

The Porter or loaded governor[76]
In the 1850s Charles T. Porter (1826–1910), of New York, turned from law, in which he had trained, to engineering. His first major invention was a stone-dressing machine, which was driven by a steam engine fitted with a conventional governor. Although the machine worked well, the stone which it dressed was found to have waves in the surface. On seeking assistance in order to correct this defect Porter was told that he should fit a larger flywheel, but convinced that the flywheel being used was adequate he 'examined the governor more critically, and made up his mind that its action was hindered by friction in the driving-joints at the top of the spindle'.[77] To reduce the forces on the driving joints, Watt-type governors were often fitted with yokes, but, as Porter argued 'The pressure applied was lighter than that applied through the joints, but it was also applied at a correspondingly increased distance from the axis, so that the effect in retarding the action of the governor was substantially the same'.[78]

Turning to his 'engineering library', which he recalls at that time consisted of Haswell's *Engineers' pocket book*, he read that 'the centrifugal force of a body revolving in any given circle varies as the square of the speed'.[79] The governor on his machine ran at 50 rev/min; thinking the matter over it occurred to him 'that if the governor could be run as fast as my machine, namely, at 300 revolutions per minute, the centrifugal force of one pound would be as great as that exerted by 36 pounds at 50 revolutions per minute'.[80]

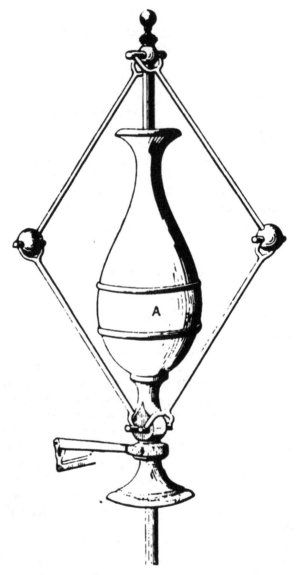

Fig. 2.12   *The first Porter loaded governor*
[Porter, C.T.: *Engineering reminiscences* (1908)]

To hold down the governor at this higher speed, an additional weight was required. In the model submitted with his patent application (1858), the weight was placed on the lever connecting the governor sleeve to the throttle valve.[81] However, it was soon to take up its familiar place between the arms of the governor, as shown in Fig. 2.12. (Fig. 2.13 shows the governor in use). The new governor reduced the oscillation of the engine so that stone with a satisfactory surface finish was produced. But the market for dressed stone was contracting, for building styles and materials were changing; Porter, therefore, abandoned the manufacture of stone-dressing machines and turned instead to the more profitable business of the manufacture of governors.

Fig. 2.13   *Porter-type loaded governor on a stationary steam engine*
[Reprinted from Watkins, G.: *The textile mill engine* David & Charles, Newton Abbot, 1970]

There is evidence that the use of weights to load the Watt governor had been tried before Porter's patent application. Weights attached to the valve linkage have been found on textile-mill engines which are known to have been installed as early as 1840[82] (it is possible, though, that the weights were a later addition). Following Porter's demonstration of his governor at the London Exhibition of 1862, a correspondent reported to *The Engineer* that he had seen at Messrs. Crawley & Sons of Newport Pagnell 'a small steam engine, the governors of which were acted upon by a heavy weight, which, by means of a notched lever, exerted any pressure which might be desired'; the Editor noted that a similar governor had long been in use at Messrs. Manning, Wardle & Co., Leeds.[83]

Charbonnier[84] in 1843 described a version of the Watt governor in which part of the restoring force was provided by a weight hung from a lever attached to the collar of the governor. The aim of this modification was not to allow the governor to be run at a higher speed, but to eliminate offset. Charbonnier realised that the ordinary governor running at nominal speed could only be in equilibrium at one height $h_0 = g/\omega_0^2$; if $h > h_0$, gravity tends to overcome centrifugal force and the balls drop; for $h < h_0$, the centrifugal force exceeds the restoring force and the balls rise. The action of the weighted lever was to apply an extra vertical force which varied with h in such a way to cancel the difference between the centrifugal force and the gravitational restoring force. The governor could therefore be in equilibrium at any height h. The required variation in restoring force was achieved by bending the lever so that the force applied to the collar increased with h.[85] Similar devices were patented by Marie Joseph Denis Farcot in 1854 and by J. B. L. Foucault in 1862.[86]

It was Porter's governor, however, which became widely used, for, as well as inventive ability, Porter also possessed commercial acumen. The shape of the weight used to load the governor is, of course, immaterial, but Porter chose to fashion it into a shape which appealed to the taste of the period. He exhibited the governor at exhibitions in London in 1862 and in Paris in 1867, visiting both exhibitions in person, and he co-operated with J. F. Allen to produce an automatic cut-off regulating gear, referred to later.

Porter was, in fact, only partially correct in his analysis of the reasons for the improvement in performance which his governor gave. Problems with the Watt governor generally occurred as a result of Coulomb friction in the governor and valve linkage. With the governor shown in Fig. 2.14*a*, the centrifugal force generated by a given change in speed can be increased by increasing the mass m of the governor balls, but this has the effects of increasing the inertia of the governor and hence the driving torque, which in turn gives rise to an increase in the friction (see Appendix 1). For an unloaded governor, the size of the dead space* due to friction is given by

$$\frac{\omega_u - \omega_l}{\omega} = \frac{F}{mg} \tag{2.2}$$

where ω_u and ω_l are the upper and lower speeds, respectively, at which the centrifugal force is sufficient to overcome the Coulomb frictional force F and g is the acceleration due to gravity.[87]

---

* This quantity was often referred to as 'detent due to friction'.

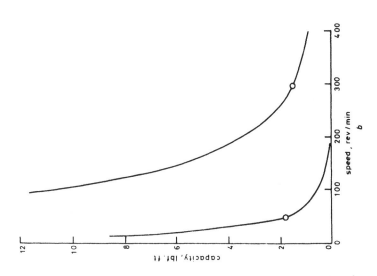

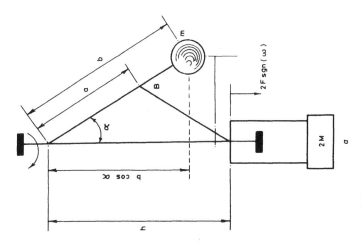

Fig. 2.14 *a* Loaded governor
b Governor capacity

An alternative way of increasing the force available to overcome friction is, of course, to increase the speed; but, since the effective moment arm is reduced when the speed of an unloaded governor is increased, the dead space remains unchanged. The effect can be seen in Fig. 2.14*b*, which shows a plot of a quantity called the 'capacity'* of the governor, and it can be seen how, as the speed is increased, the capacity decreases. By reducing the weight of the governor balls from 36 to 1 lbf, increasing the rotational speed from 50 to 300 rev/min and by using a loading weight to keep the height *h* to that obtained at 50 rev/min, Porter did not increase the capacity of his governor – as Fig. 2.14*b* shows, he actually reduced it. However, for a loaded governor, the dead space in the presence of a Coulomb frictional force *F* is given by

$$\frac{\omega_u - \omega_l}{\omega} = \frac{F}{(m + M)g} \tag{2.3}$$

and, since *F* is not directly proportional to *m* + *M* (some of the friction is in the valve linkage and is independent of *m* + *M*), the effect of loading the governor is to reduce the dead space. In fact, as was to be realised later, loading the governor enables the capacity to be increased, for no longer is the governor height solely dependent on the rotational speed (for a simple conical-pendulum governor of the Watt type, $h = g/\omega^2$); the height, and hence the capacity at a given speed, can be varied by means of the loading.

The invention of the loaded governor enabled full advantage to be taken of the use of valve cut-off gear rather than throttling for engine regulation. Throttling is an inefficient process because the throttle valve is purposely designed to cause a pressure drop and hence a loss of available energy; it is more efficient to supply the steam at constant pressure and to provide a means of changing the point at which the supply is cut off during the stroke. The benefits of this method of control had been recognised early in the century, and many engines were fitted with valve gear which enabled the cut-off point to be changed manually. In 1834 Zachariah Allen proposed the use of the Watt governor to modify the cut-off point, and similar proposals were made by Horatio Allen, J. J. Meyer and F. E. Sickels, all in 1841.

The first successful automatic cut-off valve gear was built by G. H. Corliss in 1848 and patented in 1849.[88] This mechanism used

---

* The 'capacity' of a governor is a measure of the effort or work (average force × distance moved) which can be obtained from it in response to a change of speed. It is often referred to as 'governor power', but this is a misnomer as power implies rate of doing work.

an ordinary centrifugal governor and operated on the inlet valves only. The automatic cut-off proved to be immediately successful, reducing the consumption of steam by 30 to 40%, and demand grew rapidly. It did, however, leave the basic governor mechanism unchanged. The combination of the Porter governor with the positive cut-off gear patented by J. F. Allen in 1862[89] provided a successful and widely used regulation mechanism for the developing high-speed steam engines.

Spring-loaded governors

Four years after Porter had patented the loaded governor, another American, Thomas R. Pickering, patented a spring-loaded governor.[90] A version of Pickering's governor is shown in Fig. 2.15. Instead of the traditional centrifugal pendulum, three weights supported by vertical leaf springs arranged parallel to the vertical axis of the governor are used. When small weights are used, the governor can be operated at high speeds, and the simplicity of its design, which avoids pivots and moving linkages, means that it forms an accurate and reliable governor. The Pickering governor has been widely used and is still manufactured.

A major advantage of spring-loaded governors is that they can be made physically smaller than the equivalent weight-loaded governor. Gradually they began to replace the Porter-type governor, and various designs began to be developed; Hartnell (1872)[91] and Hartung* (1870)[92] produced commercially successful models. Porter (1861)[93] designed a spring-loaded governor and, like Pecqueur (1847), Foucault (1868) and the Farcots (1864), realised that the spring loading could be adjusted to provide isochronism.[94]

It is somewhat surprising that spring-loaded governors were not developed earlier. Robert Hooke had made the proposal in 1677[95] and Marc Brunel had devised a spring-loaded governor for marine use in 1822,[96] as had John Bourne in 1834[97] and Thomas Silver in 1855.[98]

Shaft governors

In America, an alternative form of governor, the shaft governor, became popular in the 1870s. Although more complicated and expensive than the conventional governor, it had the advantage of accuracy and economy of operation. These governors are normally located within the fly wheel and employ two or more centrifugal weights arranged symmetrically and balanced by springs. The movement of the weights

* Shown in Fig. 2.16*b*.

is linked to the eccentric which operates the cut-off point in such a way that when the speed increases the throw of the eccentric is reduced and earlier cut-off is obtained.

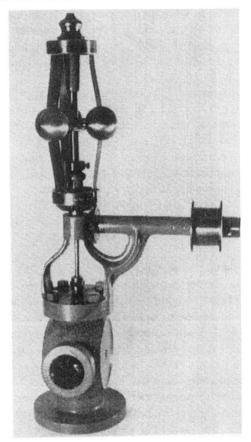

Fig. 2.15 *Pickering's spring-loaded governor*
[Reprinted by permission of the Smithsonian Institution Press from *Feedback mechanisms*, O. Mayr, *Smithsonian studies in history and technology*: number 12: Figure 32. Washington, DC: Smithsonian Institution Press, 1971]

The shaft governor was first patented in 1839 by Jacob D. Custer, but industrially successful governors did not appear until the 1870s. One such governor was that patented by Daniel A. Woodbury in 1870; it used leaf springs to restrain the weights, which were mounted to permit radial movement only. The governor was therefore purely centrifugal, as was the shaft governor designed by Wilson Hartnell and first used in 1872 (Fig. 2.16).[99]

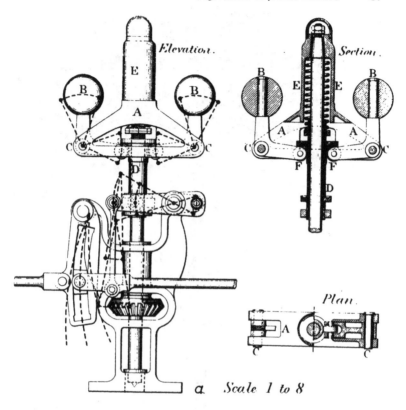

Fig. 2.16   *a* Hartnell's spring-loaded governor, 1882
              b Hartung's governor [on p. 38]
              [*Reprinted by permission of the Council of the Institution of Mechanical Engineers from Proc. I Mech. E*]

An arrangement to give purely centrifugal action is shown in Fig. 2.17*a*. By moving the pivot close to the centre of the shaft, and balancing the weight as shown in Fig. 2.17*b*, it is possible to obtain a governor which responds to inertial forces rather than centrifugal forces. The Siemens brothers had proposed governors based solely on inertial forces in 1845. Pure-inertia governors respond to the rate of change of the engine speed, i.e. the acceleration of the engine; they tend to maintain, indiscriminately, the momentary speed, but cannot be set to maintain the speed at a particular value, nor can they respond to slow changes in speed.[100] However, if the inertial action can be combined with some form of action which varies proportionately with speed, 'proportional-plus-derivative control' can be obtained. Many shaft governors, by a suitable arrangement of the pivot point (as

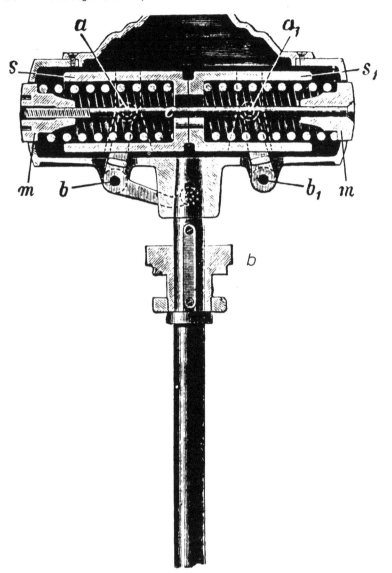

Fig. 2.16*b*

shown in Fig. 2.17*c*, for example), combined inertial and centrifugal action. Benjamin Hick's governor of 1840 combined proportional action – the force due to air resistance – with inertial action, while Throop in describing his marine-engine governor of 1857 says

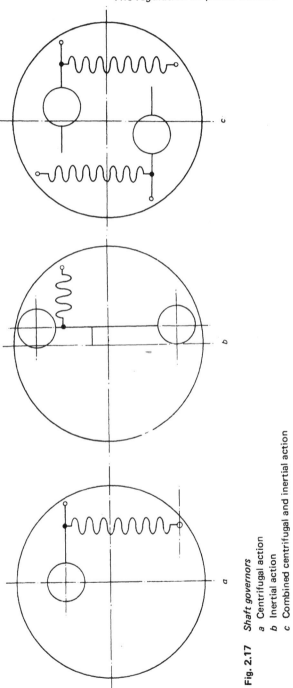

Fig. 2.17   *Shaft governors*

   a   Centrifugal action
   b   Inertial action
   c   Combined centrifugal and inertial action

> If the paddle wheel or screw ... is ... suddenly ... thrown out
> of the water ... the motion of the engine and the spindle of the
> governor is instantly increased, but the weights of the governor,
> on account of their inertia, will not readily participate in such
> increased motion, consequently they are left behind or fall back
> of the radial line, with a movement outward, aided in some degree
> by the centrifugal force due to the increased motion of the
> engine; thus instantly closing the valve.[101]

In 1865, Foucault, plagued by instability problems with his isochronous
governor, modified it so that it responded to sudden changes in acceleration as well as changes in speed.[102] But governors combining inertia
and centrifugal forces do not seem to have come into industrial use
until the 1890s.[103]

Relay governors

The traditional governor applied to the steam engine acted both as a
measuring instrument and an *actuator*. On the steam engine the various
control elements are light and well lubricated and the working medium,
steam, is elastic and of little inertia; the response to actuator movement
is fast and does not generate pressure surges. The control elements of
water wheels and turbines, however, are massive and heavy to be able
to withstand the pressure surges developed when attempts are made
to change the velocity of the water flow. Large forces are required to
operate, for example, water-wheel sluice gates, and, consequently, the
direct-acting governor cannot be used; some form of auxiliary power
is required. The governor has to be indirect acting, a relay governor.

The earliest indirect governors were purely mechanical in operation,
the governor itself operating some form of clutch to connect a shaft
or shafts driven by the water wheel to the controlling element. A good
example of this form of governor is the 'Scholfield' governor, patented
by N. Scholfield in 1836 and in 1857 in improved form.[104] A shaft
carrying two eccentrics set 180° apart is continually rotated by the
water wheel; the eccentrics operate two reciprocating arms at the ends
of which are pawls, set facing in opposite directions. The pawls engage
in ratchet wheels attached to the shaft which positions the control
gate. If the speed is correct, the governor holds both of the pawls clear
of the ratchet wheel; should the speed drop the pawl which turns the
shaft so as to open the control gate is allowed to engage its ratchet
wheel; if the speed rises, the other pawl is allowed to engage.

Improvements to this basic system, including a means of adjusting
the set speed, a backlash eliminator and a method of disengaging the
mechanism should the control sluice be fully open or fully closed, were

patented by Gibbs in 1884 and included in governors built by Rodney Hart & Co. about 1900. A much more elegant mechanical-relay governor based on the use of a friction clutch was designed by Amos and Elmer E. Woodward and described in a patent of 1890. This form of governor was manufactured by the Woodward Company around 1900. Both the Hunt and Woodward governors had a major defect in that they provided only integral action and hence the output was proportional to the time integral of the speed deviation. The governor had thus a slow response and was prone to instability. This meant that they were only suited for applications where load changes were infrequent. To correct these deficiences, the Woodward Company developed a 'compensating' governor in 1901; compensation was achieved by adding a minor feedback loop (Fig. 2.18) which used a proportion of the movement of the sluice control shaft to 'back off' the governor movement.

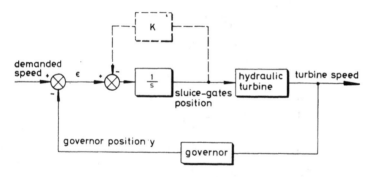

Fig. 2.18   *Block diagram of Woodward relay governor*

Mechanical systems involving clutches or points suffer from problems of wear, and mechanical servogovernors were gradually replaced by hydraulically operated relay governors. A simple form of hydraulically operated governor was patented as early as 1859,[105] but such governors did not come into general use until the end of the century. The ideas which they exploited were largely developed in connection with the provision of power steering for ships, and their development is considered in Chapter 4.

Steam-turbine regulation

Charles A. Parsons used an unusual method of regulation for his early steam turbines (Fig. 2.19). The speed-sensing device, a centrifugal fan, creates, on its suction side, a vacuum which is considered to be proportional to the turbine speed. The vacuum pressure acts on one side

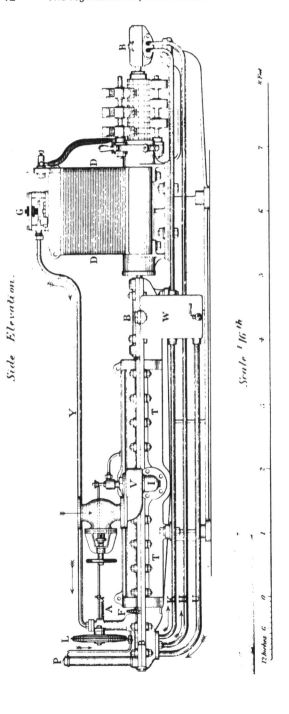

Fig. 2.19   *Parson's steam-turbine governor, 1888*
[Reprinted by permission of the Council of the Institution of Mechanical Engineers from *Proc. I Mech. E*]

of a loaded diaphragm which is connected to the throttle valve, so that, if the vacuum increases, the valve is closed. This alone would have been a rather crude system with a slow response, much inferior to the ordinary centrifugal governor; the purpose of the regulator, however, was not to control speed *per se*, but the output voltage of the generator. To achieve this, a pipe was connected to the vacuum side of the diaphragm and taken across the turbine so that its open end was above the dynamo. In front of the open end was a spring-loaded baffle pivoted so that changes in the magnetic field of the generator either closed the baffle with increasing voltage, thus increasing the vacuum and closing the throttle valve, or, with decreasing voltage, opened the baffle and consequently the throttle valve. The regulator can be considered as a pneumatic amplifier and actuator working with a variable supply pressure; as such, it forms a simple adaptive system, the sensitivity increasing with an increase in turbine speed. The overall block diagram is shown in Fig. 2.20.

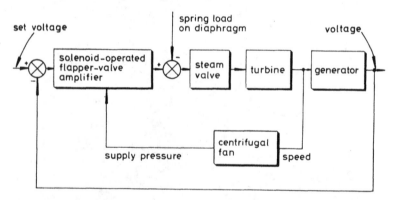

Fig. 2.20   *Block diagram of Parson's steam-turbine regulator*

The patent rights for the early turbines and regulating systems were held by Clarke, Chapman and Co., and, in 1889, when Parsons formed his own company, as well as inventing a totally new steam turbine, the radial flow turbine, he also had to find a new means of regulating it. The method he used became known as 'gust governing'; steam was admitted intermittently, i.e. in periodic gusts. By this artifice the control mechanism was kept in continual motion, thereby avoiding errors due to static friction and making the system responsive to small changes in speed. The problem of static friction was more acute in the steam turbine in that the governor was not subject to the cyclic variations in speed which occur in a steam or gas engine. Also, by

Fig. 2.21   *Parson's steam turbine with 'gust' governor*
[Reprinted by permission of the Smithsonian Institution Press from *Feedback mechanisms*, O. Mayr, *Smithsonian studies in history and technology*: number 12: Figure 52. Washington, DC: Smithsonian Institution Press, 1971]

this time, improvements in gear cutting had reduced another source of vibration.

Gust governing was implemented in several ways: in some cases the duration of the gust was directly controlled by the regulator, in others the gust action was simply obtained by superimposing motion derived from an eccentric on the normally regulated setting of the throttle valve. A turbine of the latter type is shown in Fig. 2.21. The generator output voltage is supplied to a solenoid, and the core of the solenoid, suspended from a spring, is drawn down by an amount proportional to the current in the solenoid, and this movement, by means of a lever arrangement, operates the throttle valve. The gust effect is obtained by an arrangement which oscillates the pivot point of the lever. Another notable feature of this particular engine was that the throttle movement was servoassisted, the power source being the steam which leaked through the bushings of the throttle valve.[107] The use of 'dither' (the modern term for the principle underlying gust governing) is now an accepted technique, particularly in hydraulic systems.

Conclusion

By the end of the 19th century a wide range of engine governors was available commercially. These governors were of different sizes (capacity), and graphical techniques had been developed to aid the steady-state design of governors. The problem of the regulation of prime movers had apparently been solved; gross instability could now be avoided and there was as yet no commonly available means of measuring small fluctuations in speed. Engineers were concerned with improvements in mechanical construction: reduction in size, improved relay mechanisms, methods of adjusting set speed and reduction in friction. They were not, with the exception of a small group in Germany, concerned with the dynamics of the governor. But the use of the governor by scientists, by Airy for the regulation of the motion of telescopes and by Maxwell and his colleagues in an experiment to determine the ohm, stimulated investigations of the dynamic behaviour and led to the formulation of criteria for stability.

References and notes

1 The concept of prime-mover admits no precise definition; see MOLELLO, A. P.: 'The electric motor, the telegraph, and Joseph Henry's theory of technological progress', *Proc. IEEE*, 1976, 64, p. 1275

2 MUMFORD, L.: *The myth of the machine: technics and human development* (Secker & Warburg, London, 1967)

3 BURSTALL, A. F.: *A history of mechanical engineering* (Faber & Faber, London, 1963), pp. 126–133. Mechanical clocks were predated by water clocks and, as Mayr has shown, the first feedback mechanism was part of a water clock [Mayr, O.: *The origins of feedback control* (MIT Press, Cambridge, Mass., 1970), pp. 11–16]

4 See FULLER, A. T.: 'The early development of control theory – I', *J. Dynamic Systems, Measurement & Control, Trans. ASME*, series G, 1976, 98, pp. 109–118 for details of Huygens's work. Note also the comment on this paper by BENNETT, S.: 'A note on the early development of control theory', *J. Dynamic Systems, Measurement & Control, Trans. ASME* series G, 1977, 99, pp. 211–213

5 BENNETT, S.: 'The search for "uniform and equable motion": a study of the early methods of control of the steam engine', *Int. J. Control*, 1975, 21, pp. 120–132

6 *ibid.*, pp. 132–139

7 MUIRHEAD, J. P.: *The origin and progress of the mechanical inventions of James Watt* (Murray, London, 1854 3 vols.), vol. 3, p. 37

8 *ibid.*, p. 143

9 British Patent 615, 1745, Edmund Lee, 'Self regulating wind Machine'. See MAYR: *Origins*, pp. 93–99

10 MAYR: *Origins*, p. 97

11 British Patent 1484, 1785, ROBERT HILTON, 'Windmills'. See MAYR: *Origins*, pp. 99–100 for details

12 British Patent 1628, 1787, THOMAS MEAD, 'Regulator for wind and other mills'

13 ALDERSON: *Mechanics Magazine*, 1825, 4, p. 238

14 For example, North Leverton, Nottinghamshire. The windmill has weight-loaded shutter sails, a fan tail and a lift tenter

15 See the Watt–Dearman correspondence in DICKINSON, H. W., & JENKINS, R.: *James Watt and the steam engine* (Oxford University Press, 1927), pp. 221–222

16 SMEATON, J.: *Reports* (Longman, Hurst, Rees, Orme & Brown, London, 1812 Vol. 2), p. 396

17 Watt to Boulton, 17 April 1786, quoted from Dickinson and Jenkins, *op. cit.*, p. 64

18 Boulton to Watt, 28 May 1788, quoted from Dickinson and Jenkins, *op. cit.*, p. 220

19 NICHOLSON, W.: *A journal of natural philosophy, chemistry and the arts*, 1798, Vol. 1, p. 419

20 YOUNG, T.: *Lectures on natural philosophy and the mechanical arts* (London, 1807, 2 Vols.) See MAYR: *Origins*, pp. 113–115, for details of early publications referring to the governor

21 Drinkwater to Boulton and Watt, 21 November, 1789, quoted in Dickinson and Jenkins, *op. cit.*, p. 221

22 PREUSS, J.: 'On a new steam-engine governor', *Philos. Mag.*, 1823, 62, pp. 297–299

23 HANNUIC, P.: British Patent 8623, 1840. 'Governors etc.'

24 BENNETT: *op. cit.*, p. 125. Field, in the discussion on J. Woods's paper, 'Exhibition and description of the chronometric governor invented by E. W. and C. W. Siemens', *Minutes & Proceedings of the Institution of Civil Engineers*, 1846, 5, pp. 255–265 attributes the hydraulic pump governors to Watt, and COMBES, C.: 'Rapport sur régulateur à insufflation présenté par M. L. Molinié', *Bulletin de la Société d'Encouragement pour l'Industrie Nationale*, 1841, **40**, pp. 349–367, says that the idea of a pneumatic pump regulator was imported from England

25 MAYR: *Origins*, pp. 115–118

26 The earliest reference to pump regulators avoiding offset appears to be in COMBES, *op. cit.*, p. 360

27 Information on Molinié is taken from FULLER, A. T.: 'The early development of control theory – II', *J. of Dynamic Systems, Measurement & Control, Trans. ASME*, 1976, **98**, pp. 224–235

28 COMBES: *op. cit.*, pp. 352–357. Combes refers to a similar governor having been described in the *Journal of the Franklin Institute*, March 1837, and that a similar governor had been in use at the works of M. Fourchambault for 12 years

29 *ibid.*, p. 359

30 Davison, discussion on paper by Woods: *op. cit.*, p. 265

31 British Patent 8613, 1840, BENJAMIN HICK, 'Regulators or governors for steam engines etc.'

32 British Patent 2835, 26 March 1805, 'Steam engines'

33 REES, A,: *The cyclopaedia; or universal dictionary of arts, sciences and literature*, (Longmans, London, 1819–1820)

34 Field, contribution to discussion on paper by WOODS: *op. cit.*, p. 264

35 MAYR, O.: 'Victorian physicists and speed regulation: an encounter between science and technology', *Notes and Records of the Royal Society*, 1971, **26**, pp. 205–228 for a full account

36 British Patent 10151, 1844, JOSEPH WOODS 'Regulating the power and velocity of machines for communicating power'

37 WOODS: *op. cit.*, pp. 255–265

38 *ibid.*, p. 255

39 *ibid.*, p. 255

40 *ibid.*, p. 256

41 SIEMENS, E. W.: 'Description of the differential governor of the brothers Werner and William Siemens', *The Scientific and Technical Papers of Werner von Siemens* (Murray, London, 1895), p. 12. English translation of 'Beschreibung des Differenz-Regulators der Gebrüder Werner and William Siemens zu Berlin', *Dinglers' Polytechnisches Journal*, 1845, **98**, pp. 81–89

42 *ibid.*, p. 2

43 *ibid.*, p. 6

44 SIEMENS, C. W.: 'On an improved governor for steam engines', *Proc. I Mech. E*, 1853, p. 76

45 SIEMENS, E. W.: *op. cit.*, pp. 491–495

46 Many forms of mechanical integrator were developed during the 19th century, largely for use in planimeters; see HORSBURGH, E. M.; *Handbook of the Napier Tercentenary Exhibition* (Edinburgh, 1914). Maxwell had designed an improved integrator in 1855; MAXWELL, J. C.: *Transactions of the Royal Scottish Society of Arts*, 1865, **4**, and this was followed by the

wheel–globe–cylinder integrator of James Thomson: 'On an integrating machine having a new kinetic principle', *Proc. Roy. Soc.*, 1876, **24**, p. 262. Thomson's brother, Lord Kelvin, used this type of integrator to build a harmonic analyser (THOMSON, W. [Lord Kelvin]: 'Harmonic analyzer', *Proc. Roy. Soc.*, 1878, **27**, pp. 371–373) which was put to use as a tide analyser. He also suggested how, in principle, such integrators could be used for the solution of the general 2nd-order differential equations (THOMSON, W.: 'Mechanical integration of linear equations of the second order with variable coefficients', *Proc. Roy. Soc.*, 1876, **24**, p. 269)

47 MAYR: *Notes & Records*, pp. 205–228
48 SIEMENS, C. W.: *op. cit.*, 1853, pp. 75–87, pl. 15–18
49 SIEMENS, C. W.: 'On uniform rotation', *Trans. Roy. Soc.*, 1866, **156**, pp. 657–660, pl. 29–30; see also FULLER: *Control theory – II*, pp. 15–16
50 FULLER: *Control theory – I*, p. 112
51 MAYR, O.: *Feedback mechanisms in the historical collections of the national Museum of History and Technology*, (Smithsonian Institute Press, Washington, 1971), p. 16; pp. 4–69 deal with governors
52 HEAD, J.: 'Paper on the simple construction of a steam engine governor having a close approximation to perfect action', *Proc. I. Mech. E.*, 1871, p. 213
53 For a detailed account of Foucault's work on governors see Mayr: *Notes & Records*, 1971, **26**, pp. 205–228
54 FOUCAULT, L.: *Recueil des travaux scientifiques*, Gariel, C-M. (Ed.). (Paris, 1878, 2 Vols.), vol. 1, pp. 501–503, pl. 19: 1–4
55 MAYR: *Notes & Records*, pp. 213–215
56 FULLER: *Control theory – II*, p. 225
57 *ibid.*, pp. 226–227
58 HEAD: *op. cit.*, p. 218
59 FULLER: *Control theory – II*, p. 228
60 ROLLAND, E.: 'Mémoire sur l'establissement des régulateurs de la vitesse', *Comptes Rendus*, 1867, **64**, pp. 1005–1008; Supplement *Comptes Rendus*, 1868, **66**, p. 305; also in *Journal de l'École Polytechnique*, 1870
61 de ROMILLY, W.: 'Mémoire sur divers systèmes de régulateurs à force centrifuge', *Annales des Mines*, 1872, **1**, (7th series), pp. 36–64
62 VILLARCEAU; 'Sur les régulateurs isochronous dérnés du système de Watt', *Comptes Rendus*, 1872, **74**, pp. 1437–1485
63 MARIE, G.: 'Etude comparee des régulateurs de vitesse, de pression, de temperature', *Annales des Mines*, 1878, **14**, (7th series), pp. 450–548
64 FULLER: *Control theory – I*, p. 112
65 PONCELET, J-V.: *Cours de mécanique appliquée aux machines*, Kretz, M. X., (Ed.) (Gauthier-Villars, Paris, 1874), pp. 63–123
66 TRINKS, W.: *Governors and the governing of prime movers* (Constable, London, 1919), p. 147
67 WADDELL, R.: 'On an escape-water-valve, and a governor for marine steam engines', *Proc. I Mech. E.*, 1853, pp. 118–120, pl. 27
68 JENSEN, P.: 'On a marine engine governor', *ibid.*, 1859, pp. 92–95, pl. 23
69 BELL, R. B.: Discussion, *Proc. ICE* 1877–1878, **51**, p. 34
70 *The Engineer*, 1859, **8**, pp. 95, 155
71 MAYR: *Feedback mechanisms*, pp. 18–19
72 *ibid.*, p. 19

73 BALL, F. H.: 'Steam engine governors', *Trans. ASME*, 1897, **18**, p. 291

74 THOMPSON, S. P.: *Dynamo-electric machinery: a manual for students of electrotechnics* (E. & F. N. Spon, London, 1892), p. 596

75 WEBB, C. R., and JANATA, M. S.: 'Governors with load sensing', *Proc. of the I Mech. E.*, 1971, **184**, Pt. 1, pp. 161–180

76 See MAYR: *Feedback mechanisms*, pp. 13–16, and PORTER, C. T.: *Engineering reminiscences* (Wiley, New York, 1908)

77 PORTER: *op. cit.*, p. 18

78 *ibid.*, p. 18

79 *ibid.*, p. 19

80 *ibid.*, p. 19

81 MAYR: *Feedback mechanisms*, p. 14

82 WATKINS, G.: *The textile mill engine* (David & Charles, Newton Abbot, 1970), Vol. 1, see Figures 1, 2, 5, 6 and 7

83 *The Engineer*, 1863, **15**, p. 97

84 CHARBONNIER: 'Mémoire sur les moyens généralement employés pour régulariser les mouvements des machines à vapeur à manivelles, et en particulier sur le régulateur employé par MM. J.-J. Meyer et Comp.', *Bulletin de la Société Industrielle de Mulhouse*, 1843, **17**, pp. 332–385

85 FULLER: *Control theory – II*, p. 225

86 *ibid.*, pp. 227–228

87 See, for example, GREEN, W. G.: *Theory of machines* (Blackie, London, 1955), pp. 466–471

88 MAYR: *Feedback mechanisms*, p. 8

89 *ibid.*, p. 14

90 MAYR: *Feedback mechanisms*, pp. 28–31

91 HARTNELL, W.: 'Automatic expansion gear', *Proc. I Mech. E.*, 1882, **33**, pp. 408–430

92 See HALL, H. R.: *Governors and governing mechanisms* (The Technical Publishing Co. Ltd., Manchester, 1907) for details

93 PORTER: *op cit.*, p. 34

94 FULLER: *Control theory – II*, p. 227

95 HOOKE, R.: *Lampas: or description of some mechanical improvements of lamps and waterpoises together with some other physical and mechanical discoveries* (Royal Society, London, 1677). Reprinted in GUNTHER, R. T.: *Early science in Oxford* (Oxford University Press, 1931)

96 ROLT, L. T. C.: *Isambard Kingdom Brunel* (Penguin Books, Harmondsworth, 1970), pp. 323–324

97 BOURNE, J.: *The steam engine* (Artisan Club, London, 1846)

98 MAYR: *Feedback mechanisms*, p. 18. Silver's governor was used on steamships of the British and French navies

99 HARTNELL: *op. cit.*, p. 419, Pls. 74–76

100 Many purely inertial governors are described in 19th century patent and other literature; most make use of, in some form or other, a loosely coupled flywheel

101 Quoted from MAYR: *Feedback mechanisms*, p. 19

102 FULLER: *Control theory – II*, p. 228

103 For further information on shaft governors see Mayr: *Feedback mechanisms*, pp. 33–38

104 This governor was still being manufactured in 1900, see MAYR: *Feedback mechanisms*, p. 39

105 *The Engineer*, 1869, 7, p. 370
106 TRINKS: *op. cit.*, p. 171
107 MAYR: *Feedback mechanisms*, p. 48

Chapter 3

Towards an understanding of the stability of motion

There is scarcely any question in dynamics more important for Natural Philosophy than the stability or instability of motion.

W. Thomson and P. G. Tait, *Treatise on natural philosophy*, 1867

Introduction

As we observed in the previous chapter the development of the flyball governor was the outcome of a letter written by Matthew Boulton to James Watt. Although Watt was quick to recognise from the information given in the letter that the lift tenter could be adapted to regulate the steam engine, it seems doubtful that he ever recognised the full potential of his governor, that is, the dynamic nature of its operation. Boulton implied in his letter that the important feature of the lift tenter was that it allowed the engine to start under a light load: 'I think the principal advantage of this invention is in making it easy to set the engine to work because the top stone cannot press upon the lower until the mill is in full motion'.[1] Watt apparently shared this view. In describing the governor he says that [Watt's italics] 'the *Governor* . . . shuts it [the throttle valve] more or less according to the speed of the engine, so that as the velocity augments, the valve is shut, until the speed of the engine and the opening of the valve come to a maximum and balance each other'.[2] The behaviour is conceived in terms of equilibrium; on starting up, or on making a major change to the load by, say, the addition of another pair of stones, the governor would make the necessary change to the position of the throttle valve, but then would, in effect, be inoperative. There is no conception of dynamics, of the possibility of overshoot, of the governor being constantly in action to correct disturbances.

That Watt should describe the behaviour of the governor in this manner is not surprising; prevailing notions were still founded on statics, perfect balance: Dalton's description of the steam engine ran

> High on huge axis heav'd above,
> See balanc'd beams unweary'd move[3]

But ideas were changing. Mayr has shown that political economists – in particular, David Hume and Adam Smith – were beginning to think in terms of dynamic balance.[4] In *An inquiry into the nature and causes of the wealth of nations,* Adam Smith wrote,

> The natural price is the central price, to which the prices of all commodities are continually gravitating. Different accidents may sometimes keep them suspended a good deal above it, and sometimes force them down even somewhat below it. But whatever may be the obstacles which hinder them from settling in this centre of repose and continuance, they are constantly tending towards it.[5]

If there are disturbances, there will be 'occasional and temporary fluctuations in the market price of any commodity' as the system approaches equilibrium.

John Robison, Professor of Natural Philosophy at the University of Edinburgh, writing to Watt in 1783 said, 'I consider machines as in motion, performing work. It is evident that this view must lead to or require very different maxims of construction from those which result from the equilibrium of machines, the point of view in which they have generally been considered.'[6] And again, writing to Watt in 1796, regarding his proposed book, *Elements of mechanical philosophy,* Robison wrote [the italics are his], 'All this [*sic*] in a way totally different from the usual treatises of equilibrium of Forces – but in a machine *doing work* this equilibrium is destroyed in one sense, while it remains perfect in another sense, viz the Equilibrium between the forces *actually* impelling and the resistance *actually* exerted'.[7]

It was not until 38 years after its invention that a clear description of the dynamic behaviour of the governor was given. John Farey, in his book on the steam engine published in 1827, describes the action of the governor and throttle valve in the following way [the italics are his]:

> The effect of this apparatus to regulate the velocity of an engine is not quite perfect ... The revolving pendulums cannot operate beforehand to prevent alterations of velocity, but only to correct the alternations after they have taken place; and *such corrections*

will always be in excess; for whenever the governor has altered the throttle-valve to a sufficient extent to correct any sudden change in the motion of the engine, it must retain the valve in its altered position, until the *motion of the engine is affected thereby*, and the centrifugal force of the governor being also affected, it will produce an alteration of the throttle-valve, in a contrary direction to the first one; another alteration of speed must therefore ensue, and then a consequent correction will be made by the governor and throttle-valve. As each succeeding correction becomes more and more minute, the motion of the engine is not perceptibly deranged thereby, . . . providing that the governor is properly proportioned; . . . whenever a great and sudden change of the motion of the engine takes place, the governor will make a correspondingly great and sudden alteration of the opening of the throttle-valve, so as not merely to correct the change of the motion, but it will cause an alteration of an opposite character to that of the original one, though to a less extent. The governor will afterwards apply a remedy to that second alteration, and proportionably less active, though still in excess; and thus, after a few fluctuations the governor will bring the valve to its proper adjustment. To prevent any evil from this property of the governor, it must be made to act delicately on the throttle-valve, so that a considerable expansion or contraction of the pendulum will not produce too much alteration to the opening of the valve.[8]

In this description there is a clear understanding that the governor requires time to act, that it cannot 'prevent alterations in velocity'; an understanding that overshoot can occur because of the time required to act and that the degree of overshoot can be controlled by adjusting the relationship between the movement of the governor and the opening of the throttle-valve – in modern terms, by reducing the loop gain. Farey continues, 'The office of the governor is to correct small and casual derangements of the *dynamic equilibrium* of the forces of the machine, which may arise from fluctuations of the resistance which is opposed by the mill-work, or from variations in the force of the engine',[9] leaving us in no doubt that the governor was now being considered as a 'machine doing work'.

The tenuous nature of the 'dynamic equilibrium' of the governor was made clear by Airy, who in 1840 wrote, 'Whenever the equilibrium of forces requires that a free body be brought to a determinate position, we must not expect that the body will remain steadily in that position of equilibrium, but that it will oscillate on both sides of that position, and that . . . it will have no tendency to settle itself in the position of equilibrium'.[10]

Another contribution by Farey was a description of the causes of offset:

A different quantity of steam will be required for every permanent alteration of the resistance, but the governor can make no change in the opening of the throttle-valve, except in consequence of a change of the velocity of the engine; and that ought to be avoided, by adjusting the connexion between the governor and the throttle-valve according to every permanent change of resistance.[11]

As was noted in the previous chapter, much effort was devoted to the elimination of offset, and the consequence of many of the modifications made was to increase the tendency to hunt. In this chapter we are concerned with the growing awareness of the dynamic behaviour of the governor, the attempts to analyse this behaviour, and, above all, attempts to understand the nature of dynamic stability.

The discovery of dynamics

Steady-state analysis

Thomas Young, writing of the Watt governor in 1807,[12] limited his analysis to the formula for the height h of an ideal conical pendulum rotating at speed ω, which he gave as

$$h = g/\omega^2 \tag{3.1}$$

A more extensive treatment was undertaken by Jean-Victor Poncelet, who, although not concerned with the kinetic behaviour of the governor (he did not use differential equations to describe the behaviour), was aware of the importance of Coulomb friction. In the first edition of his *Cours de Mécanique* (1826)[13] he derived the equilibrium equation for a governor of the type shown in Fig. 3.1a. In the 1836 edition a more extensive treatment is given and he obtains the following expression showing the effect of Coulomb friction on speed regulation:

$$\frac{P}{p} = \frac{a}{b} \frac{\omega^2}{\omega_u^2 - \omega^2} \tag{3.2}$$

$P = mg$ (m = mass of the governor balls), p is the friction force, ω_u is the upper speed and ω is the nominal speed.

Assuming small changes of speed, eqn. 3.2 can be rearranged to give

$$\frac{\omega_u - \omega_l}{\omega} = \frac{a}{b} \frac{p}{P} \tag{3.3}$$

where ω_l is the lower speed, $\omega = (\omega_u + \omega_l)/2$. This quantity became known as the 'coefficient of sensibility',[14] 'degree of irregularity'[15] or 'detent due to friction'.[16]

Poncelet considered two other forms of governor construction

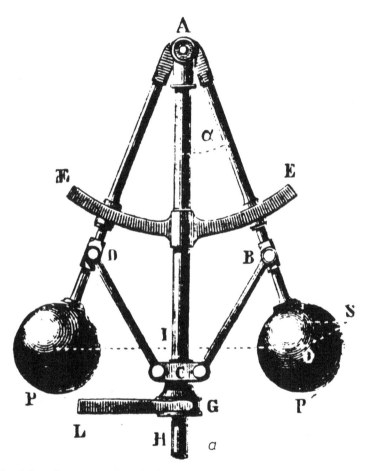

Fig. 3.1a   *Governor analysed by Poncelet*

(Fig. 3.1*b* and *c*) and assessed the affect of friction on them. He shows that they both reduce the effect of Coulomb friction, but at the same time reduce the change of height of the conical pendulum for a given change of speed. He concludes that, in practice, the governor balls should not exceed 40 kg, noting that, if this does not supply sufficient force to operate the control valve when using a governor of the form shown in Fig. 3.1*a*, one of the other forms of governor construction (Fig. 3.1*b* or *c*) should be adopted.

Because of its limited circulation, Poncelet's work seems to have had little influence on the theoretical or practical development of the governor. It was not until the end of the 19th century that a similar clear and systematic treatment of the governor appeared.

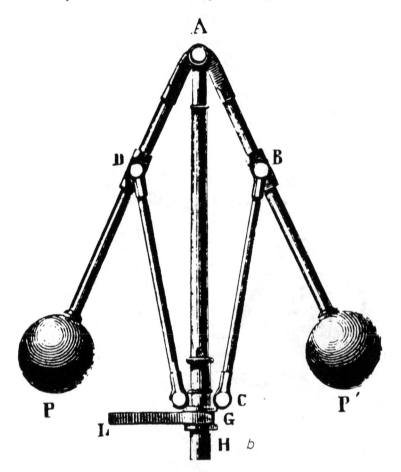

Fig. 3.1*b*   *Governor analysed by Poncelet*

G. B. Airy and the investigation of dynamic stability

> ... he Airy was the most absolutely original person with whom I ever came into contact ... the whole of the observatory [Greenwich] was full of his inventions – doors which shut by contrivances of his own, arrangements for holding papers, for

making clocks go simultaneously, for regulating pendulums, for arranging garden beds, for keeping planks from twisting, for every conceivable thing from the greatest to the smallest. On all there was the impress of an original and versatile mind, bubbling over with inventiveness . . .

James Stuart, *Reminiscences,* 1912

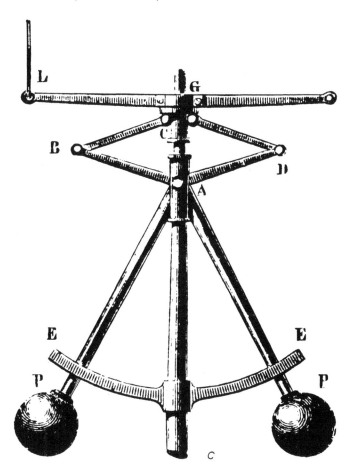

Fig. 3.1c   *Governor analysed by Poncelet*

Sir George Biddell Airy (1801–1892), Astronomer Royal from 1835–1881, was a man of immense talent and enormous activity, publishing over 500 papers and 11 books. During the greater part of the 19th century his patronage was vital to the aspiring scientist or engineer: he sat in judgment on such matters as the design of the Britannia

Bridge, the strength of the Atlantic cable, railway gauges, Babbage's computers, sewer systems and magnetic compasses in iron ships. But, as Mayr has pertinently remarked, these serve only to demonstrate Airy's interest, not his competence, in technology.[17] He is also remembered for some spectacular misjudgements, such as his condemnation of Babbage's calculating engine, his failure to find the new planet (Neptune) and his erroneous estimate of the wind loading on the Forth Bridge.

Airy's interest in friction governors arose because of the need, in making astronomical observations, to keep the telescope rotating slowly at a uniform speed round its polar axis to compensate for the earth's rotation. If this is not done, the star being observed moves out of the field of view in a few seconds. Hooke described an apparatus for rotating a telescope at constant speed in 1674,[18] but the first practical application was not made until 1824, when Fraunhofer constructed a large telescope at Dorpat which was driven by clockwork, the speed being regulated by a centrifugal governor.[19] There was a crucial difference between Fraunhofer's and Hooke's use of the centrifugal pendulum: Hooke had relied on air resistance to provide the force resisting overspeed, Fraunhofer made use of the changes in centrifugal force to vary the friction opposing the motion of the governors.

Airy described the regulator in the following way:

> the axis of the regulator is vertical; it carries a horizontal cross-arm, to the extremities of which are attached springs nearly transverse in direction to the cross-arm, carrying at their ends small weights; a drum surrounds the regulators: when the regulator is made to revolve with a certain velocity, the centrifugal force of the balls bends the springs till the balls just touch the inner surface of the drum, the smallest additional velocity causes the balls to press against the drum, and creates such a friction as will immediately reduce the velocity to the determinate velocity at which the balls will just touch the drum . . .[20]

Similar governors were used on telescopes in England by both Sheepshanks and Airy in 1834.

The friction governor tends to give rise to small-amplitude high-frequency oscillations which would be smoothed out by a steam engine, but on a telescope the observer would see the oscillation magnified by the optical system. Experience with friction governors brought this problem to the attention of Airy, and, in a paper written in 1840, he attempted to investigate it mathematically.[21] The paper begins by drawing attention to the dynamic nature of the instability:

Whenever the equilibrium of forces requires that a free body be brought to a determinate position, ... we must not expect that the body will remain steadily in that position of equilibrium, but that ... it will have no tendency to settle itself in the position of equilibrium: and we must take account of this possible oscillation in planning any mechanism ... In practice there always exists one cause which tends constantly to reduce the oscillations ... namely ... friction ... [22]

The main body of the paper can be divided into three sections, analysing:

(a) the behaviour of the conical pendulum itself, ignoring driving forces, friction and inertia of the telescope
(b) the effect of modifying the governor by (i) adding a flywheel to the vertical axle (this is equivalent to taking into account the referred inertia of the telescope) and (ii) by suspending the governor balls from a crossarm
(c) the behaviour of the governor under the action of a driving force.

In section (a), dealing with the basic conical pendulum, as shown in Fig. 3.2a, Airy derives a nonlinear differential equation of first order, but second degree,

$$\left(\frac{d\theta}{dt}\right)^2 + \frac{k_3}{S^2\theta} - \frac{2g}{a}\cos\theta = k_4 \tag{3.4}$$

Attempting to linearise this equation, he considers small deviations in θ, i.e. he assumes that

$$\theta = \alpha + \zeta(t) \tag{3.5}$$

Unfortunately, substituting for θ in eqn. 3.4 gives, neglecting squares of small quantities, an algebraic equation

$$k_5 = k_6\zeta = 0 \tag{3.6}$$

in which k_5 and k_6 are constants. Since the only time-varying quantity is ζ, k_5 and k_6 must be zero. To avoid this difficulty, Airy retains the ζ^2 terms and hence obtains an equation

$$\left(\frac{d\zeta}{dt}\right)^2 + k_5 + k_6\zeta + k_7\zeta^2 = 0 \tag{3.7}$$

where k_7 is a constant. He now assumes that, since ζ is small and since the conical pendulum is a conservative system, the solution to eqn. 3.7 will be of the form

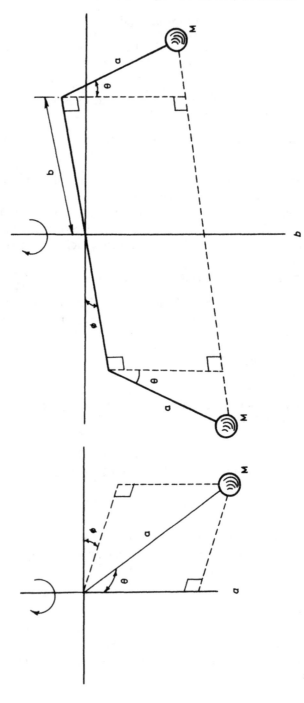

Fig. 3.2   *Governor arrangements analysed by Airy in 1840*

$$\zeta = \gamma \sin{(mt + n)} \tag{3.8}$$

where γ, m and n are constants.

On the basis of this assumption Airy finds that, if t is the time of a revolution of the pendulum about the vertical axis and t' is the period of oscillation,

$$\frac{t'}{t} = (4 - 3 \sin^2{\alpha})^{-1/2} \tag{3.9}$$

Hence it follows that the period of the oscillation lies between the times of a half revolution and a full revolution.

Using the same technique, Airy shows that, if a flywheel is added to the system, the period of the oscillation can be made large by making the inertia of the flywheel large. He also found that, by placing the suspension rods at the end of a crossarm (Fig. 3.2b), the period of the oscillation can be made small by increasing the length b of the crossarm.

In considering the forced motion of the system by including the effects of the driving force and friction force (in effect closing the feedback loop), Airy, for simplicity, restricts the analysis to an ordinary governor without flywheel or crossarm. He derives the differential equation for the system and, as above, restricts attention to small changes ζ of θ from its nomial value α. Since the system now being considered is nonconservative, we no longer expect a sinusoidal solution, but one of the form

$$\zeta = e^{pt+q} \sin{(mt + n)} \tag{3.10}$$

Airy was able to show that p is positive, thus concluding that 'the amplitude of the oscillation of the balls . . . goes on perpetually increasing; and will become larger until the law of friction becomes different from that assumed in the investigation'.[23]

Although not achieving any practical solution to the problem of oscillation of the friction governor, Airy drew attention to the problem of instability of dynamic systems and showed how the instability of the system could be accounted for by a consideration of the differential equations of the system. His difficulties arose because he tried to proceed with the linearisation of the differential equation obtained directly from the angular-momentum and energy equations. This procedure gives rise to an equation (eqn. 3.4) of first order, but second degree, which cannot easily be linearised. Had he seen that by differentiating eqn. 3.4 to give

$$\frac{d^2\theta}{dt^2} - k_3 \frac{\cos\theta}{\sin^3\theta} + \frac{g}{a}\sin\theta = 0 \qquad (3.11)$$

linearisation about the operating point becomes simple and the whole analysis would have been much less cumbersome.

Some years later Airy returned to the problem. He had become an enthusiastic supporter of Siemens's chronometric governor; he used it on a new telescope at the Liverpool Observatory, on a chronograph recorder and on the Great Equatoreal [*sic*] telescope at Greenwich in 1846.[24] The Siemens's governor, although an improvement on the basic governor, was prone to give rise to high-frequency oscillations, which caused a blurring of the image when the governor was applied to the telescope. Airy investigated these oscillations and showed that they could be reduced by means of a hydraulic damper.

He published his findings in 1851 in the form of a supplement to his paper of 1840.[25] In it he writes [the italics are his],

If we can indicate any part of the mechanism which must have an *absolute* oscillation connected with the *relative* oscillation which we wish to check, [i.e. can some part of the mechanism be found which does not rotate with the pendulum but still shows the oscillation due to the instability] and if we subject the motion of the indicated part of the effects of friction (using the word in the most general sense, as expressing force always opposed to the direction of motion) of sufficient magnitude, we shall certainly destroy both the absolute oscillation and the relative oscillation; and if the law of the friction be such that, when the velocity is indefinitely diminished, the friction also is indefinitely diminished, then the absolute oscillation and the relative oscillation will be suppressed, without interfering in the smallest degree with the legitimate action of that indicated part.[26]

Airy states that he knows of only one form of friction which satisfies the above conditions, that which arises from the movement of a plate through water or another perfect fluid. If the frictional force is proportional to velocity, the equation of motion

will always have the form

$$\frac{d^2x}{dt^2} + a\frac{dx}{dt} + bx = 0 \qquad [3.12]$$

a and *b* being positive; and the solution of this equation will be one of the three following:

$$x = A.e^{-mt} \cos (nt + B)$$
$$x = (At + B)e^{-mt} \qquad\qquad [3.13]$$
$$x = A.e^{-mt} + Be^{-mt}$$

either of which denotes a very rapidly diminishing displacement ... [27]

He does not say what he means by x in eqn. 3.12. If he means system error, the difference between the angular velocity and the desired value, he is incorrect, since the total system is third order, as shown in Fig. 3.2b. It would, however, seem that Airy had gone back to considering the free motion of the conical pendulum plus damper, rather than the overall speed-control system.

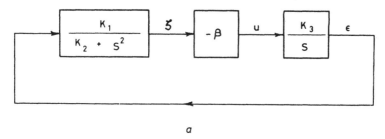

a

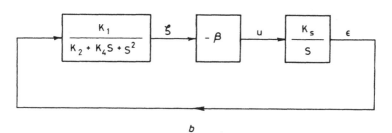

b

Fig. 3.3   *Block diagrams of governor mechanisms analysed by Airy*
     a In 1840
     b In 1851

The significance of this supplement is that Airy had arrived at a method of stabilising the governor; using block-diagram notation, the system he analysed in 1840 can be shown as in Fig. 3.3a, and the effect of his 1851 proposals is to change the system to that shown in Fig. 3.3b, which, as the accompanying Nyquist plot of Fig. 3.4 shows, can be made stable.

J. C. Maxwell on governors

In 1861 the British Association for the Advancement of Science appointed a committee to establish electrical standards. William Thomson, Clerk Maxwell and William Siemens were members of the committee, and Thomson designed the experiment for the determination of the ohm. Crucial to this experiment was the rotation of a coil

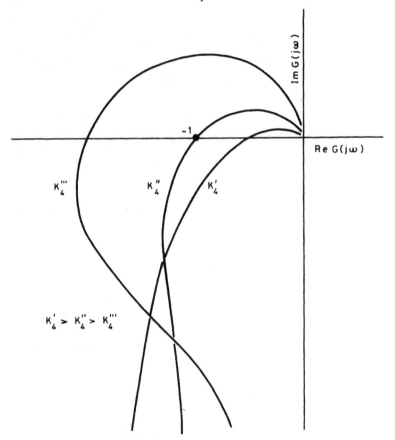

Fig. 3.4   *Nyquist plot of system analysed by Airy in 1851, showing the effect of the damping term*

at a constant speed, and a governor for the purpose was designed by Henry Charles Fleeming Jenkin (1833–1885). The experiment itself was carried out in 1863 at King's College, the University of London, by Maxwell, Balfour Stewart and Jenkin.[28] No contemporary description of the governor has been found, but the governor itself is preserved in the Whipple Museum of Science, Cambridge University. The principle of its operation is illustrated in Fig. 3.5. It is basically a friction governor; if the speed increases, the force between the balls and the friction ring increases, but the friction ring, instead of being held stationary, can rotate and raise up a weight (if the speed decreases, the weight causes the ring to rotate in the opposite direction). The weight is hung in a cylinder of water to provide hydraulic damping.

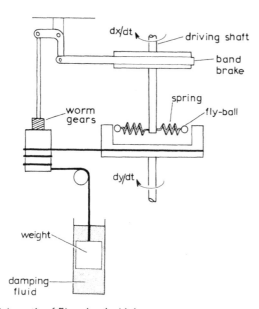

Fig. 3.5 *Schematic of Fleeming Jenkin's governor*

V_1 = nominal velocity
F = Coulomb friction
B = moment of inertia of auxiliary wheel
Y = viscous-damping coefficient
W = constant torque due to weight
M = moment of inertia of machine
G = constant relating torque applied to machine to movement y
P = driving torque
R = load torque

The movement of the ring is used to slacken or tighten a band brake which acts on the main driving shaft; the system therefore includes power amplification. The governor also provides integral action: if the speed varies from the nominal, a torque is applied to the horizontal wheel, which will turn at a rate determined by the torque and the viscous friction in the damper; the movement of the band brake is given by the integral of the rate of rotation of the wheel.

Through this work both William Thomson and Clerk Maxwell became interested in engine governors. William Thomson in 1866 read a paper to the Glasgow Philosophical Society in which he described the electrically regulated chronometers of R. J. James and A. Bain.[29] He also devised a chronometer of his own, which he described in 1869;[30] details of refinements to it were published in 1876.[31] The governor used in this chronometer was patented[32] and offered for general use. Thomson also designed a simple mechanical friction governor for use in telegraphic apparatus and chronographs,[33] which, by means of a simple modification suggested by Fleeming Jenkin, could also be used to regulate the steam engine.[34]

Maxwell's interest in the governor was, however, largely theoretical; he was concerned with the governor as an example of a *dynamic system*. He had at an earlier period of his life addressed himself to the problem of dynamic stability when, in 1856, he won the Adams Prize for an essay on the stability of Saturn's rings.[35] In this essay he drew attention to 'a very general and very important problem in Dynamics', which he expressed in the following way: 'Having found a particular solution of the equations of motion of any material system, to determine whether a slight disturbance of the motion indicated by the solution would cause a small periodic variation, or a total derangement of the motion'.[36] He notes that this problem can be solved in principle by the application of the calculus of variations, but that, when applied to a function of several variables, will be so intricate that he is 'doubtful whether the physical or the abstract problems will be first solved'.[37]

Maxwell's approach to the problem of Saturn's rings was to approximate and linearise the equations; by doing so he avoided the problems of solving nonlinear differential equations which had hindered Airy's work on dynamic stability. He was aided by a stroke of luck. The characteristic equation corresponding to the linearised differential equations of motion was biquadratic and hence could be easily factorised to obtain the location of the roots. He was thus able to concentrate on discussing the nature of the behaviour which would result from various locations of the roots and in showing which locations would give stability.

During this period interest in dynamic stability was growing. William Thomson and P. G. Tait included a long section on stability in their book *A treatise on natural philosophy*[38] published in 1867. Their approach to the problem – probably devised by Tait – [39]was based on Hamiltonian dynamics: a general principle of stability was outlined in terms of the *principle of varying action*. In analysing the general dissipative system, they observed that the nature of the behaviour of the system was determined by the roots of a certain determinantal equation, the equation we now call the characteristic equation. On extending the analysis to an 'artificial or imaginary cumulative system; they remark that 'If the roots ... of the determinantal equation ... are all real and negative, the equilibrium is stable: in every other case it is unstable'.[40] There was no attempt to determine the location of the roots from an examination of the coefficients of the equation.

Maxwell, a close friend of both Thomson and Tait, knew of the book and probably knew of the approach being used, but did not see the book until travelling to London at the beginning of 1868.[41] He presented papers to a meeting of the London Mathematical Society on 23rd January and, in the course of the discussion on another paper (presented by a Mr. J. J. Walker), Maxwell asked, 'if any member present could point out a method of determining in what cases all the possible [real] parts of the impossible [complex] roots of an equation are negative'. For he had, 'In studying the motion of certain governors for regulating machinery ... found that the stability of the motion depended on this condition, which is easily obtained for a cubic, but becomes more difficult in the higher degrees'.[42] It was in reply to this request that W. K. Clifford made the statement 'that by forming an equation whose roots are the sums of the roots of the original equation taken in pairs and determining the condition of the real roots of this equation being negative we should obtain the condition required'.[43]

This suggestion of Clifford's is in principle a solution to the problem of finding a general stability criterion. Given a polynomial equation with real coefficients,

$$f_1(x) = p_0 x^n + p_1 x^{n-1} + \ldots + p_n \tag{3.14}$$

the roots will be of the form $\alpha, \beta \pm \gamma i$. Taking the sum of the roots two by two gives an auxiliary equation

$$f_2(x) = P_0 x^m + P_1 x^{m-1} + \ldots + P_m \tag{3.15}$$

with roots $\alpha + \beta \pm \gamma i, \beta$.

The complex roots of $f_1(x)$, since they must occur in conjugate pairs, give rise to real roots in $f_2(x)$. Therefore, by using the well known rule of signs (Descartes), $f_1(x)$ and $f_2(x)$ can be tested for positive real roots. If neither has a positive real root the system, with a characteristic equation corresponding to $f_1(x)$, is stable.

Clifford's suggestion thus reduced the problem to that of

determining the coefficients $P_0, \ldots, P_m$ of $f_2(x)$ in terms of the coefficients $p_0, \ldots, p_n$ of $f_1(x)$. The techniques for the determination of coefficients of an equation whose roots were a given function of the roots of another equation had been widely studied and were well known. However, the determination of the auxiliary equation by this method rapidly becomes unwieldy as the degree of $f_1(x)$ increases. Maxwell, however, does not seem to have made use of Clifford's suggestion; there is no reference to it in his paper 'On governors', which was received by the Royal Society on 20th February 1868.[44]

This paper, probably the most widely known paper on automatic control, opens with Maxwell carefully distinguishing between regulators which give offset, which, following Siemens, he terms *moderators*,[45] and *governors* which have no offset. Later in the paper he is less precise and sometimes uses 'governor' to describe any form of speed controller. After declaring that it is his intention 'to direct the attention of engineers and mathematicians to the dynamical theory of such governors', he describes the modes of behaviour of a machine with a governor:

> It will be seen that the motion of a machine with its governor consists in general of a uniform motion, combined with a disturbance which may be expressed as the sum of several component motions. These components may be of four different kinds:
>
> 1. The disturbance may continually increase.
> 2. It may continually diminish.
> 3. It may be an oscillation of continually increasing amplitude.
> 4. It may be an oscillation of continually decreasing amplitude.
>
> The first and third cases are evidently inconsistent with the stability of motion; and the second and fourth alone are admissible in a good governor. This condition is mathematically equivalent to the condition that all the possible [real] roots, and all the possible [real] parts of the impossible [complex] roots, of a certain equation shall be negative.[46]

Since he was not 'able completely to determine these conditions for equations of a higher degree than the third', Maxwell expressed the hope that the subject would gain the attention of the mathematicians. There then follows a paragraph in which Maxwell explains the limitations of the static methods of design and the possible utility of his approach [Maxwell's italics] :

> The actual motions corresponding to these impossible [complex] roots are not generally taken notice of by the inventors of such machines, who naturally confine their attention to the way in which it is *designed* to act; and this is generally expressed by the

real root of the equation. If, by altering the adjustments of the machine, its governing power is continually increased, there is generally a limit at which the disturbance, instead of subsiding more rapidly, becomes an oscillating and jerking motion, increasing in violence till it reaches the limit of action of the governor. This takes place when the possible [real] part of one of the impossible [complex] roots becomes positive. The mathematical investigation of the motion may be rendered practically useful by pointing out the remedy for these disturbances.

In the analysis which follows, Maxwell mentions the controllers of Watt, Fleeming Jenkin, William Thomson, Foucault, Airy and William Siemens. The descriptions given are, however, fragmentary; he assumes that his readers are familiar with these controllers.[48] Differential equations describing various systems of increasing complexity are obtained. For the governor of Fleeming Jenkin (Fig. 3.5) he obtains an equation of motion for the auxiliary wheel:

$$B\frac{d^2y}{dt^2} = F\left(\frac{dx}{dt} - V_1\right) - Y\frac{dy}{dt} - W \tag{3.16}$$

and, for the machine itself,

$$M\frac{d^2x}{dt^2} = P - R - F\frac{dx}{dt} - V_1 - Gy \tag{3.17}$$

where the symbols are as shown in the block diagram of Fig. 3.6*a*.

For small changes in angular velocity, dx/dt, since V_1 and W are constant, the system reduces to that shown in Fig. 3.6*b*. Maxwell obtained for this system the characteristic equation

$$MBn^3 + (MY + FB)n^2 + FYn + FG = 0 \tag{3.18}$$

He notes that one root of this equation is evidently real and negative (since it is of odd degree with positive coefficients), and then writes that

The condition that the real part of the other roots should be negative is

$$\left(\frac{F}{M} + \frac{Y}{B}\right)\frac{Y}{B} - \frac{G}{B} = \text{a positive quantity.}$$

This is the condition of stability of the motion. If it is not fulfilled there will be a dancing motion of the governor, which will increase till it was as great as the limits of motion of the governor. To ensure this stability, the value of Y must be made sufficiently great, as compared with G, by placing the weight W in a viscous liquid if the viscosity of the lubricating materials at the axle is not sufficient.[49]

Maxwell gives no proof of the stability criterion. There is, however, no difficulty in obtaining this criterion for a 3rd-order system; Fuller has suggested an approach which Maxwell may well have used.[50]

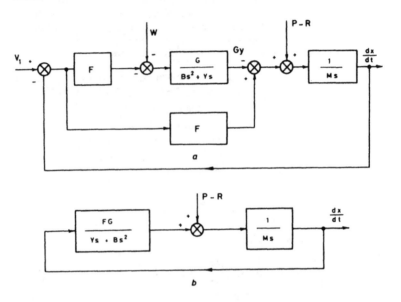

Fig. 3.6   *Block diagram of Fleeming Jenkin's governor*

The analysis of Thomson's and Foucault's governors also gives rise to a characteristic equation of third order. To obtain a system of higher order Maxwell conjectured a governor in which the 'break of Thomson's governor is applied to a moveable wheel, as in Jenkin's governor'.[51] The nature of this governor is not made clear,[52] but it leads Maxwell to a 5th-order system with a characteristic equation which he writes as

$$n^5 + pn^4 + qn^3 + qn^2 + sn + t = 0 \qquad (3.19)$$

Two *necessary* conditions for stability, he asserts, are that $pq > r$ and $ps > t$, but he says, 'I am not able to show that these conditions are sufficient'.[53]

The paper 'On governors' is terse and at times enigmatic; it lacks Maxwell's usual lucidity. Fuller has suggested that it was probably written in a forced attempt to free his mind from control problems and thereby allow him to concentrate on the writing of his treatise on electricity and magnetism.[54] This lack of clarity, coupled with the lack of analysis of commonly used governors (although Maxwell's approach, in terms of energy, was applicable to a wide range of governors), meant that the paper could not be readily understood by the majority of contemporary engineers and scientists.

The basic ideas were, however, quickly taken up by E. J. Routh, who wrote a new section on governors for the second edition of his book *Rigid dynamics,* published in 1868. And over the next 50 years the various editions of this book were largely responsible for disseminating the idea of, and methods of assessing, dynamic stability.

Continental developments

Writing in 1859, Franz Reuleaux (1829–1905), the great German kinematicist,[55] besides distinguishing between 'static' and 'astatic' governors, classified governors as being 'tachometric' or 'dynamometric', the latter being feedforward devices, their action depending on the measurement of the load. Two years later J. Luders analysed the steady-state behaviour of simple governors, obtaining as Poncelet had done, an expression for 'the degree of irregularity' (see p. 55).[56] In 1865, Luders attempted to consider the dynamics of engine governors, but succeeded only in obtaining the changes in kinetic energy in terms of the work input and the work output.[57]

In 1871 L. Kargl, lecturer, and later Professor, at the Polytechnic in Zurich, obtained differential equations for the motion of governors of the Watt and Porter types; he included the effects of Coulomb friction in the analysis, and, although unable to obtain an analytic solution for the equations, he determined the transient response from step-by-step solution.[58] In a second paper two years later, Kargl showed the impracticability of a purely astatic governor.[59]

The most important European contribution came, however, from a Russian engineer J. Wischnegradski.[60] His analysis of the governor was first reported to the Académie des Sciences in Paris on 31st July 1876, by M. Tresca. It was published in Russia in 1877 and translated into German in the same year; in 1878 and 1879 the full version appeared in French.[61] The text of the paper has never appeared in English.[62]

Wischnegradski considered the combination of a steam engine and governor. Unlike Maxwell, he explains his assumptions and the development of the differential equation in detail. The system considered is shown in block-diagram form in Fig. 3.7. By assuming small changes and neglecting the Coulomb friction, Wischnegradski obtained the equation

$$\frac{du^3}{dt} + \frac{Md^2u}{dt^2} + \frac{Ndu}{dt} + \frac{KgL}{I\omega_0} u = \frac{Kg}{I\omega_0}(p - Q)\rho \qquad [3.20]$$

of which he says, 'L'équation $(b)^*$ qui détermine la nature du u en fonction du temps t, est une équation linéaire du troisième ordre à coefficients constants; son intégrale dépendra, comme l'on sait, des racines de l'equation:

$$\theta^3 + M\theta^2 + N\theta + \frac{KLg}{I\omega_0} = 0 \, .$$ [3.21]

Pour abréger le discours, on donnera à cette équation l'épithè de *caractéristique* . . .'[63]

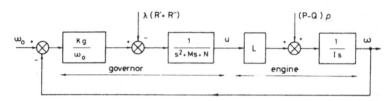

Fig. 3.7   *Block diagram of engine and governor analysed by Wischnegradski*

ω_0	= nominal speed
Kg	= centrifugal-force constant
$\lambda(R' + R'')$	= Coulomb-friction term
$(P - Q)g$	= disturbance
M	= constant depending on viscous friction
N	= restoring-force constant
u	= movement of throttle valve
L	= torque constant
I	= moment of inertia of engine and load

There then follows an extensive discussion of the behaviour of the governor according to the nature of the roots of the characteristic equation: all roots real and unequal; two equal roots; two imaginary roots with negative real parts; two imaginary roots with positive real parts etc. By substituting

$$\theta = \phi \sqrt[3]{\frac{KLg}{I\omega_0}}, \quad M = x \sqrt[3]{\frac{KLg}{I\omega_0}}, \quad N = y \sqrt[3]{\left(\frac{KLg}{I\omega_0}\right)^2}$$

in eqn. 3.21, Wischnegradski obtains an equation

$$\phi^3 + x\phi^2 + y\phi + 1 = 0$$ (3.22)

with two parameters x and y, which have become known in Russian and German literature as the Wischnegradski parameters.[64] On the basis of these two parameters, Wischnegradski drew a diagram showing

---

* Eqn. 3.20

the location of the roots, in effect a stability diagram, for a 3rd-order system.[65] This diagram is shown in Fig. 3.8. It divides into three regions each region corresponding to a different type of system behaviour:

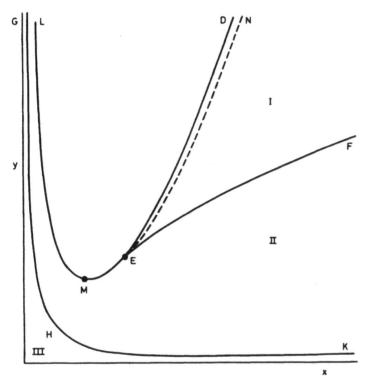

Fig. 3.8   *Stability diagram from Wischnegradski*

(*a*) Bounded by LME and EF, in which following a disturbance, the regulator moves to a new position without oscillation.

(*b*) Bounded by LMEF and GHK, in which, following a disturbance, the movement is oscillatory, but the oscillation dies out.

(*c*) Bounded by the axis and GHK, in which, following a disturbance, the movement is oscillatory, the oscillation continuing to increase.[66]

The curve GHK is a rectangular hyperbola

$$xy = 1 \tag{3.23}$$

and hence the condition for stability is

$$xy > 1 \qquad (3.24)$$

Substituting for x and y gives

$$MN\frac{I\omega_0}{KLg} > 1 \qquad (3.25)$$

where, as Wischnegradski showed, $N\omega_0/KLg = $ slope of the speed/torque curve for the steam engine. This expression clearly shows, as Wischnegradski states in his conclusions, that isochronous (astatic) regulators, because they cause instability, cannot be used with success.

Wischnegradski's attempt was no closer to a general solution of the problem of dynamic stability than Maxwell's attempt had been nine years earlier; it was, however, of greater significance to the engineer wishing to design a governor. The derivation of differential equations is clearly explained and the stability requirements are presented in a useful and easily understood form. It was particularly important that a German translation was produced,[67] for during the last quarter of the 19th century and the early part of the 20th century the major developments in the study of the dynamics of machinery were taking place in Germany.[68]

Stability and governor design

E. J. Routh
Edward John Routh, born on 20th January 1831 in Quebec, Canada, gained the coveted title of Senior Wrangler in the 1854 mathematical tripos at Cambridge (Maxwell was Second Wrangler). After graduating, Routh began work as a private tutor in mathematics at Peterhouse, his Cambridge college, and between 1855–1888 he coached more than 600 undergraduates, 27 of whom became Senior Wranglers.[69] In addition to his academic connections, Routh was also linked to the group of leading scientists through marriage, for in 1864 he married Hilda, Airy's daughter.[70]

The stability condition for a 3rd-order system given by Maxwell in his 1868 paper was, as we have noted, incorporated by Routh into the second edition of his book. A demonstration of the derivation of the condition for the equation

$$ax^3 + bx^2 + cx + d = 0 \qquad (3.26)$$

with roots $x = \alpha \pm \beta i, \gamma$, using the well known relationships between the coefficients of an algebraic equation and its roots,

$$-\frac{b}{a} = 2\alpha + \gamma \tag{3.27a}$$

$$\frac{c}{a} = 2\gamma\alpha + \alpha^2 + \beta^2 \tag{3.27b}$$

$$-\frac{d}{a} = (\alpha^2 + \beta^2)\gamma \tag{3.27c}$$

was given. From eqns. 3.27*a, b* and *c* it can easily be shown that

$$\frac{bc - ad}{a^2} = -2\alpha((\alpha + \beta)^2 + \gamma^2). \tag{3.28}$$

Hence $(bc - ad)/a^2$ will always have the opposite sign to α; for α to be negative, therefore,

$$\frac{bc - ad}{a^2} > 0 \tag{3.29}$$

which is one of Maxwell's conditions, the other being that *a, b, c* and *d* should all have the same sign.

Six years later, in 1874, Routh returned to the problem and, by making use of Clifford's suggestion made to Maxwell at the beginning of 1868, was able to determine a complete set of stability criteria for 5th-order system.[71]

In 1875 the notice giving the subject for the Adams Prize, to be adjudged in 1877, was issued. One of the examiners was Maxwell, and the subject was 'The criterion of dynamical stability'. Routh submitted his essay 'A treatise on the stability of a given state of motion' in 1876. He was awarded the prize, and the essay was published unchanged in 1877.[72] The essay, 108 pages long and divided into 8 chapters, is Routh's most innovative work. It is in Chapter III of this work that the so-called Routh conditions for stability are derived.

After defining such terms as small quantity, stable motion, steady motion and explaining how the nature of the motion of a body is determined by the roots of a certain 'determinantal' characteristic equation, Routh turns, in Chapters II and III, to considering ways of determining stability without finding the roots of the characteristic equation.

His first approach is based on Clifford's suggestion of forming an auxiliary equation $F(D) = 0$ whose roots are the sums of the roots of the characteristic equation $f(D) = 0$ taken in pairs, which leads him to state [the italics are his], *'Our first test of the stability of a dynamical system is that all the coefficients of the dynamical equation f(D) = 0 and all the coefficients of its derived equation F(D) = 0 should have the same sign'.*[73] For an equation of order n this test gives rise to $\frac{1}{2}n(n+1)$ inequalities, which are, as Routh points out, all necessary and sufficient, but are not all independent. In pursuing this approach he gives several methods of determining the coefficients of the auxiliary equation, and notes for particular examples how the number of inequalities considered can be reduced.

However, Routh abandons this approach and turns to a theorem given by Cauchy regarding a polynomial equation

$$f(z) = p_0 z^n + p_1 z^{n-1} + \ldots p_{n-1} z + p_n = 0 \qquad (3.30)$$

He states theorem in the following way:

> Let $z = x + y\sqrt{-1}$ be any root, and let us regard x and y as coordinates of a point referred to rectangular axes. Substitute for z and let
>
> $$f(z) = P + Q\sqrt{-1} \qquad [3.31]$$
>
> Let any point whose coordinates are such that P and Q both vanish be called a radical point. Describe any contour, and let a point move round this contour in the positive direction and notice how often P/Q passes through the value zero and changes its sign. Suppose it changes α times from $+$ to $-$ and β times from $-$ to $+$. Then Cauchy asserts that the number of radical points within the contour is $\frac{1}{2}(\alpha - \beta)$. It is, however, necessary that no radical point should lie *on* the contour.[74]

This therorem is a special case of Cauchy's index theorem.[75] In using the theorem Routh first considers the behaviour of the function when the infinite semicircle bounding the positive half of the xy plane is traversed from $-\infty$ to $+\infty$. He follows Sturm in expressing $f(z) = 0$ in polar co-ordinates:

$$f(z) = p_0 r^n (\cos n\theta + \sin n\theta \times \sqrt{-1}) + \ldots \qquad (3.32)$$

and hence

$$P = p_0 r^n \cos n\theta + p_1 r^{n-1} \cos (n-1)\theta + \ldots \qquad (3.33)$$

$$Q = p_0 r^n \sin n\theta + p_1 r^{n-1} \sin (n-1)\theta + \ldots \qquad (3.34)$$

Since r is infinite, in the limit

$$P/Q = \cot n\theta \qquad (3.35)$$

and therefore P/Q will vanish when

$$\theta = \frac{2k+1}{n} \frac{\pi}{2} \quad (k \text{ an integer}) \qquad (3.36)$$

Routh thereby deduces that, if n is even, the sign of $P/Q = \cot n\theta$ will change from positive to negative n times, and, if n is odd, $n - 1$ times (the angle θ is assumed to be in the interval $-\pi/2 < \theta < \pi/2$) in traversing the infinite semicircle from $-\infty$ to $+\infty$. He then argues from Cauchy's theorem that for there to be no roots of $f(z) = 0$ in the right-hand halfplane the sign of P/Q should change from $-$ to $+$ in traversing the y-axis as many times as it changed from $+$ to $-$ when

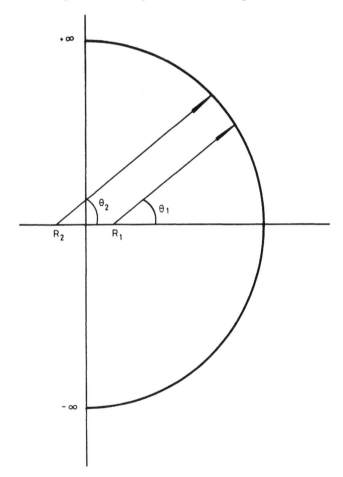

Fig. 3.9   *Contour of integration in the complex plane*

traversing the semicircle. That this is so can be seen from Fig. 3.9. For roots in the right-hand halfplane such as R_1 the angle θ continues to increase as the y-axis is traversed and hence P/Q will pass through a zero from positive to negative, whereas for R_2, θ is decreasing as the y-axis is traversed and hence when P/Q passes through a zero it does so from negative to positive.

The condition for stability has therefore been changed into a requirement that, in traversing the y-axis from $+\infty$ to $-\infty$, P/Q should change sign from negative to positive n or $n-1$ times, according to whether n is even or odd. Routh states this requirement in the following way [the italics are his]:

> to express the necessary and sufficient conditions that $f(z) = 0$ *may have no radical point on the positive side of the axis of y, put* $z = y\sqrt{-1}$ *and equate to zero separately the real and imaginary parts. Of the two equations thus formed, the roots of the one of lower dimensions must separate the roots of the other. It is also necessary that the coefficients of the two highest powers of z in f(z) should have the same sign.*[76]

The equations formed by this procedure are

$$P(y) = p_n - p_{n-2}y^2 + p_{n-4}y^4 - \ldots = 0 \tag{3.37}$$

$$Q(y) = p_{n-1}y - p_{n-3}y^3 + p_{n-5}y^5 - \ldots = 0 \tag{3.38}$$

The problem has now been changed into one of finding a method 'to express in an analytical form the conditions that the roots of an equation $f_2(x) = 0$ may be all real, and may separate the roots of another equation $f_1(x) = 0$ of one degree higher dimensions.'[77]

In solving this problem Routh makes use of a theorem given by Sturm in 1835.[78] Fourier (1820)[79] and Budan (1807) had derived results concerning the maximum number of real roots of an equation $f_0(x) = 0$ in a given interval, which, when all the roots are known to be real, gives the number of roots in the interval. The method required the formation of the series of derived functions of $f_0(x)$,

$$f'(x), f''(x), f'''(x), \ldots, f^{(n)}(x)$$

In 1829 Sturm generalised Fourier's rule so as to determine the number of real roots in the interval $A < x < B$ even if $f(x) = 0$ possessed complex roots. Then in 1835 he gave a rule which can be expressed in the following way:[80]

From the given polynomial equations

$$f_0(x) = a_0x^n + a_1x^{n-1} + \ldots + a_n = 0 \quad (a_0 \neq 0) \tag{3.39}$$

$$f_1(x) = b_0x^m + b_1x^{m-1} + \ldots + b_m = 0 \quad (m < n) \tag{3.40}$$

form the sequence of functions

$$f_0(x), f_1(x), f_2(x), \ldots, f_r(x)$$

where

$$f_1(x) = f_0'(x)$$

$$f_2(x) = -\,[\text{remainder obtained on dividing } f_0(x) \text{ by } f_1(x)]$$

$$f_3(x) = -\,[\text{remainder obtained on dividing } f_1(x) \text{ by } f_2(x)]$$

Continue this process until a remainder $(-f_{r+1})$ is obtained which is identically zero. Then, when x increases in the interval $A < x < B$, the number of real roots of $f_0(x) = 0$ for which f_0 and f_1 changes from unlike to like signs, minus the number of real roots of $f_0(x) = 0$ for which f_0 and f_1 change from like to unlike signs, is equal to the number of variations of sign in the sequence

$$f_0(A), f_1(A), \ldots, f_r(A)$$

minus the number of variations of sign in the sequence

$$f_0(B), f_1(B), \ldots, f_r(B)$$

Multiple roots are only counted once and any zero terms are omitted from the sequence.

This rule is not directly what Routh required, but Sturm in developing it gives the rule

the excess ϵ of the number of times when the quantity V/V_1 $[f_0(x)/f_1(x)]$ or P/Q in vanishing for different points of the line AB [part of contour in the xy plane] passes from positive to negative, over the number of times when it passes from negative to positive, will be equal to the excess of the number of variations which are found in the sequence of signs of the functions V, $V_1, \ldots, V_r$ for $S = \beta$, [at B] over the number of variations for $S = \alpha$ at A.

Now, if the roots of $f_1(x)$ separate those of $f_0(x)$, the sign of P/Q will change alternately as $f_0(x)$ and $f_1(x)$ vanish, and hence the direction of change of sign when $f_0(x)$ vanishes is always the same. If P and Q start with the same sign, Sturm's rule will, in this case, give a gain in the number of variations in sign; if P and Q have opposite signs at A, a loss in the number of variations in sign will be indicated. This gain or loss in the number of variations in sign should be n if the required conditions relating to $f_0(x)$ and $f_1(x)$ are to be satisfied.

In the particular case of dynamic stability where

$$f_0(x) = P(x) = p_n - p_{n-2}x^2 + p_{n-4}x^4 - \ldots = 0 \quad (3.41)$$

$$f_1(x) = Q(x) = p_{n-1}x - p_{n-3}x^3 + p_{n-5}x^5 - \ldots = 0 \quad (3.42)$$

the function are alternately of even and odd degree, and hence Routh was able to express the necessary and sufficient conditions for stability in the form that 'The coefficients of the highest powers of x in the series

$$f_1(x), f_2(x), f_3(x) \ldots$$

must all have the same sign'.[82]

Having determined these conditions Routh notes that the process of finding the Sturmian functions, i.e. the process of finding the greatest common measure [divisor] is long, but it may be shortened, he says, by omitting the quotients and performing the division as follows [the italics are his] :

Let

$$f_1(x) = p_0 x^n - p_2 x^{n-2} + p_4 x^{n-4} - \ldots$$
$$f_2(x) = p_1 x^{n-1} - p_3 x^{n-3} + p_5 x^{n-5} - \ldots$$

[3.43]

then, since p_1 is positive, it easily follows by division that

$$f_3(x) = A x^{n-2} - A' x^{n-4} + A'' x^{n-6} - \ldots$$

[3.44]

where

$$A = p_1 p_2 - p_0 p_3$$
$$A' = p_1 p_4 - p_0 p_5$$

etc. = etc.,

so that by remembering this simple cross-multiplication *we may write down the value of $f_3(x)$ without any other process than what may be performed by simple inspection.* In the same way $f_4(x)$ etc. may all be written down.[83]

Observing that, although only the first terms of each of the expressions $f_1(x), f_2(x), \ldots, f_n(x)$ are required, all the terms have to be found, he gives his second algorithm, which says that each succeeding term $f_i(x)$ can be obtained from the previous term $f_{i-1}(x)$ by making the following substitutions: for

$$p_0, p_1, p_2, p_3, \ldots$$

write[84]

$$p_1, p_1 p_2 - p_0 p_3, p_3, p_1 p_4 - p_0 p_5 \ldots$$

Routh's second approach to the problem of stability, based on Cauchy's theorem of the argument and Sturmian division, is an improvement on Clifford's proposal in that it involves only n inequalities, rather than $\frac{1}{2} n(n + 1)$.

Up to this point Routh had made two reservations: (*a*) that there were no roots on the imaginary axis, and (*b*) that none of the functions

$f_1(x), f_2(x), \ldots, f_n(x)$ vanished absolutely. Examining these two cases he observed that they can be considered as one and he showed that a vanishing term $f_i(x)$ could be replaced by the differential coefficient of the previous term $f_{i-1}(x)$.[85] In ending the chapter Routh considers as an example the Watt governor

The influence of Routh on governor design

Routh gradually incorporated the formulas for the stability conditions into the various editions of his book on rigid dynamics.[86] Knowledge of the conditions did not, for many years, spread beyond the scientific circle of Maxwell, Thomson, Airy, Siemens and others. Ordinary engineers remained unaware of this theoretical work. In England the gap between theory and practice was large throughout the 19th century. In 1846 Bourne commented that 'Tredgold's work is chiefly made up of mathematical sublimities, which have but little relation to practice'.[87] By the late 1850s engineers were beginning to use Tredgold's formula for the height of a conical pendulum; they were, however, concerned as to how the height was to be measured – was it from the point of suspension or from the point of intersection of a line drawn through the centre of the balls and the point of suspension with the vertical axis? Controversy on the point continued in the columns of *The Engineer* for several years.[88]

The practising engineer of the 1880s who ventured into Routh's books was not likely to be persuaded to study the theory when he read that 'A common defect of governors is that they act too quickly, and thus produce considerable oscillation of speed in the engine. . . . This fault may be very much modified by applying some resistance to the motion of the Governor'.[89] By this time the action of dashpots in reducing the governor oscillation was well known, but it was also known that such action gave rise to sluggish systems.

On reading further they would find Routh considering another defect of the common governor, namely offset: 'In the case of Watt's Governor if any permanent change be made in the relation between the driving power and the load, the state of uniform motion which the engine will finally assume is different from that which it had before the change. . . . This defect may be considerably decreased by the use of Huygen's parabolic pendulum'.[90] For twenty-five years engineers had, without success, been trying to use the parabolic pendulum, but as William Siemens said in 1871, 'such an action would not be practically applicable to regulate the speed of a steam engine, because, as had been explained, there would be no tendency for the balls to stop in their movement if the governor were chronometric, but they would fly at once from one extremity of their range to the other under the slightest alternations of speed'.[91]

To use the parabolic pendulum as a practical governor its isochronism had to be destroyed, the usual method being the addition of a controlling spring which gave it the advantages possessed by the governors of Porter, Pickering and Hartnell, namely an increase in the force available to overcome static friction. Engineers had recognised that the governor could not be considered in isolation: 'it was clear that the throttle-valve now described in connection with the Allen governor must be looked upon as an essential part of the governor, the prompt action of the governor depending upon the ease with which the valve could be moved with a slight amount of force.[92] Whereas Routh declared, 'In this investigation no notice has been taken of the frictions at the hinge and at the mechanical appliances of the Governor, which may not be inconsiderable. These in many cases tend to reduce the oscillations and keep it without bound'.[93] They could also, of course, give rise to 'hunting'.

The above passages summarise the strength and the weaknesses of the work of Maxwell and Routh. The ability to simplify, to linearise the equations, to neglect the inessential, led to a major theoretical advance. But Maxwell had been concerned with special governors, made to high standards in which the friction was low and hence could be neglected. The typical engine governor and valve linkage was not made to such exacting standards, was often neglected, and friction forces were of the essence.

By the time Routh's work became widely available in the 1880s the problem of violent oscillation of engine governors had largely been solved by the adoption of loaded governors (weight or spring loaded) and by careful design of the valve linkage. Hunting was not entirely eliminated but it was reduced to a level which could be tolerated. The designer was now looking for a method which would enable him to determine the necessary size of governor to give a specified steady-state accuracy.

Governor design techniques

In 1882 Wilson Hartnell, a Leeds engineer, provided the foundations for the steady-state design of governors, and also gave some hints on the avoidance of hunting. Hartnell's paper[94] was based on 10 years' experience of designing spring-loaded governors of a type which he had invented in 1872. He introduced a graphical method of design, shown in Fig. 3.10. The dotted lines represent the centrifugal force due to the rotation of the governor, while the solid curved lines represent the restoring force: the curve $C_1C_3CC_2$ is the force in the absence of friction, and $F_1F_3FF_2$ and $G_1G_3GG_2$ represent the forces in the presence of friction. CF is the friction force to be overcome to move the balls outwards when the governor is at radius R, and CG is equal to that to be overcome in moving the balls inwards. On the basis of this diagram Hartnell defined the terms:

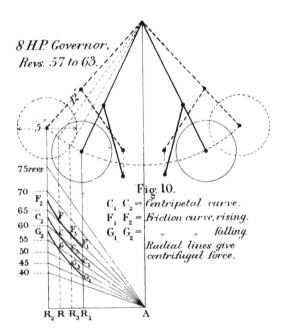

8 H.P. Governor,
Revs. 57 to 63.

75revs

Fig. 10.

C_1 C_2 = *Centripetal curve.*
F_1 F_2 = *Friction curve, rising.*
G_1 G_2 = „   „   *falling.*
Radial lines give
centrifugal force.

R_2 R R_3R_1 A

Fig. 3.10 *Hartnell's design chart*
[Reprinted by permission of the Council of the Institution of Mechanical Engineers from *Proc. I Mech.*]

governor power[95] $=$ area of $R_1C_1C_2R_2$

sensitiveness $V = \dfrac{\omega_{C2} - \omega_{C1}}{\omega_{C2} + \omega_{C1}} \times 100\%$

retarded sensitiveness $= \dfrac{\omega_{F2} - \omega_{G1}}{\omega_{F2} + \omega_{G1}} \times 100\%$

detention $= \dfrac{\omega_F - \omega_G}{\omega_C}$

For good governing Hartnell recommends that the governor power (capacity) should be at least 20 times the mean friction to be over-come; if the capacity is inadequate, continuous 'hunting' is likely to result. Recommendations regarding selection of springs, calculation of a suitable size of flywheel and choice of bell crank angle for Hartnell-type governor are also given. A similar graphical method was developed in Germany by Hermann (1886) and Tolle (1895–6).[96]

Although all the design calculations given are in terms of steady-state values, Hartnell also expresses concern about the dynamic behaviour of the engine and governor. He claims that the advantages of cut-off governing largely arises 'owing to its freedom from two evils, which may be termed "retardation from storage" and "retardation from friction"'.[97] 'Retardation from storage' is due to the steam stored between the throttle valve and the steam port. If control is by means of adjusting steam admission this effect does not occur, while the friction with the automatic cut-off is much reduced because the cut-off linkage is continually in motion. Hartnell illustrates his arguments by means of the diagrams shown in Fig. 3.11a, in which he sketched out the time response of the engine and governor under various assumptions. These diagrams show that he had a clear understanding of the dynamics involved.

The Federal Polytechnic of Zurich

In 1892 at the age of 33 Aurel Boleslaw Stodola was appointed as a professor at the Federal Polytechnic in Zurich.[98] After his appointment he devoted his attention to control problems; his first study was con-cerned with the dynamics involved in the regulation of a high-pressure water turbine, the results of which were published in 1893.[99] As part of this work Stodola solved a 3rd-order differential equation and he also used the Wischnegradski criteria to test for stability. By linearising with respect to a steady-state operating point and by introducing nondimensional quantities, he considerably simplified the mathematical presentation and he introduced into the analysis the idea of 'time constants' which was an important step towards the detachment of the 'system equations' from the technical realisation of the system thus described.

The study was extended to include the effects of the inertia and damping of the controller as well as the delayed actions of the servo-motor. As a consequence, the order of the system increased from three to seven. Stodola, at this time unaware of the work of Maxwell and Routh, guessed that the question of stability could probably be answered without the complete solution of the differential equations: he consequently presented the problem to his colleague, the

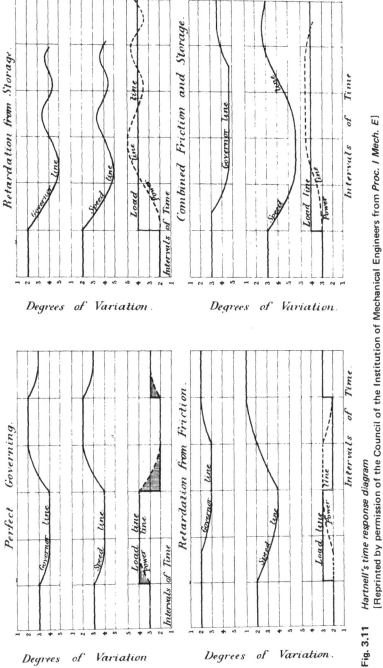

Fig. 3.11 Hartnell's time response diagram
[Reprinted by permission of the Council of the Institution of Mechanical Engineers from *Proc. I Mech. E*]

mathematician Adolf Hurwitz, who published a solution in 1895.[100]

Hurwitz, also unaware of the work of Routh and Maxwell, argues from the geometrical representation of complex numbers. He considers the entire rational function $f(x)$ of degree n and shows that if x ranges over purely imaginary values from $+i\infty$ to $-i\infty$, then N and P, the number of zeros of $f(x)$ with negative real parts and positive real parts, respectively, are given by

$$N = \frac{n+\Delta}{2}, \quad P = \frac{n-\Delta}{2} \tag{3.45}$$

Thus, for stability, $\Delta = n$. Δ is in fact the number of changes in sign of $\tan\theta$, where θ is the angle shown in Fig. 3.9. Hurwitz then argues that Δ can be obtained by the determination of a Cauchy index for a function $R(z)$. If $x = -iz$,

$$f(x) = f(-iz) = u + iv \tag{3.46}$$

then

$$R(z) = v/u \tag{3.47}$$

Note that, if

$$f(x) = p_0 x^n + p_1 x^{n-1} + \ldots + p_{n-1} x + p_n = 0 \tag{3.48}$$

then

$$R(z) = \frac{p_1 z^{n-1} - p_3 z^{n-3} + \ldots}{p_0 z^n - p_2 z^n + \ldots} \tag{3.49}$$

Up to this point Hurwitz's approach has been roughly similar to Routh's, since

$$R(z) = \frac{Q(y)}{P(y)} = \tan n\theta \tag{3.50}$$

where $Q(y)$ and $P(y)$ are given by eqns. 3.37 and 3.38 and θ is the angle defined by eqn. 3.31. From this point Routh proceeds by means of Sturm's division algorithm, but Hurwitz, while noting Sturm's algorithm, proceeds 'by the Hermitian reduction to a quadratic form'[101] to the statement of his theorem:

A necessary and sufficient condition that the equation

$$a_0 x^n + a_1 x^{n-1} + \ldots + a_n = 0$$

with real coefficients in which the coefficient a_0 is assumed to be positive, have only roots with negative real parts, is that the values of the determinants

$$\Delta_1 = a_1, \Delta_2, \Delta_3, \ldots, \Delta_n$$

all be positive.[102]

The determinants are formed as follows:

$$\Delta\lambda = \begin{vmatrix} a_1, & a_3, & a_5, & \ldots, & a_{2\lambda-1} \\ a_0, & a_2, & a_4, & \ldots, & a_{2\lambda-2} \\ 0, & a_1, & a_3, & \ldots, & a_{2\lambda-3} \\ \cdot & \cdot & \cdot & & \cdot \\ \cdot & \cdot & \cdot & & a_\lambda \end{vmatrix} \qquad (3.51)$$

$\lambda = 1, 2, \ldots, n; a_k = 0$ if $k < 0$ or $k < n$

The equivalence of the stability criteria of Hurwitz and Routh was demonstrated by Bompiani in 1911.[103]

Hurwitz passed his results to Stodola prior to publication and he was able to add a footnote to his paper saying, 'These results were applied at the Davos Spa Turbine Plant with brilliant success.[104] This work, carried out in 1894,[105] is probably the first use of stability conditions in the design of a practical control system. Stodola continued his work in control systems with an investigation of the so-called inertia controllers: shaft governors combining inertial and centrifugal action.[106] He seems to have been the first person to realise that the inertia controller was a *proportional + derivative* controller.[107]

Stodola's methods formed the basis of the extensive treatment of existing ideas of control theory given in the book by Tolle, *Regelung der Kraftmaschinen*, published in 1905,[108] the most important book on the subject of control for almost two decades.

Conclusion

The ideas and theoretical approach developed by Wischnegradski, Stodola, Hurwitz and the German-speaking engineers gradually filtered through to the English-speaking world, largely through the strong contacts which had been built up between the developing American universities and Germany. The techniques used for design were collated by Trinks and discussed in his book *Governors and the governing of prime movers*, published in 1919.[109] In terms of stability analysis, Trinks is no further advanced than Wischnegradski; he simply states the conditions of stability for a 3rd-order system.[110] As late as 1940, R. B. Smith is regretting the absence of stability analysis in American technical literature and he points out that, following on from Tolle, Stodola's work was extended by Stein (1928) and Freudenreich (1929);

the first contribution published in English being by Schwender and Luoma in 1936.[111]

In England the stability criteria developed by Routh as a consequence of Maxwell's study of governors were not used by governor designers or mechanical engineers; the first paper dealing with the dynamic analysis was not published in England until 1940,[112] and then the reference was to Tolle's book.[113] It seems to have been engineers concerned with the new inventions, the aeroplane and electrical machinery, who first recognised the value of Routh's work. In the year of the first successful flight, 1903, G. H. Bryan and W. E. Williams, analysing the longitudinal stability of an aeroplane, wrote, 'In order that the steady motion may be stable, the roots of this biquadratic $[A\lambda^4 + b\lambda^3 + C\lambda^2 + D\lambda + E = 0]$ must either be real and negative, or complex with their real parts negative and Routh finds that this will be the case if the six quantities A, B, C, D, E and $BCD - AD^2 - EB^2$ are all of the same sign'.[114] In a paper read on the same day, 18th June 1903, concerning the 'hunting' of alternating-current machinery, Bertram Hopkinson also made use of Routh's stability conditions for a biquadratic equation, referring to the 5th edition of *Rigid dynamics* (Part II).[115] From this time onwards until the 1940s, occasional use was made of the Routh–Hurwitz inequalities in testing for stability, but use did not become widespread until after the Second World War.

Theoretical work on the methods of determining the location of roots continued. As we have previously mentioned, Bompiani showed in 1911 that the conditions of Routh and Hurwitz were equivalent,[116] while in the previous year Orlando had proved that the Hurwitz inequalities were both necessary and sufficient.[117] For a polynomial with real coefficients

$$f(z) = a_0 z^n + a_1 z^{n-1} + \ldots + a_n \quad (a_0 > 0) \tag{3.52}$$

the Routh–Hurwitz conditions for stability require that n inequalities

$$\Delta_1 > 0, \quad \Delta_2 > 0, \ldots, \Delta_n > 0$$

be satisfied, where

$$\Delta_i = \begin{vmatrix} a_1 & a_3 & a_5 & \cdot & \cdot & \cdot \\ a_0 & a_2 & a_4 & \cdot & \cdot & \cdot \\ 0 & a_1 & a_3 & \cdot & \cdot & \cdot \\ 0 & a_0 & a_2 & a_4 & & \\ & & & \cdot & & \\ & & & & \cdot & \\ & & & & & a_i \end{vmatrix} \quad (a_k = 0 \text{ for } k > n)$$

is the Hurwitz determinant of order i, $i = 1, 2, \ldots, n$. In 1914 two French mathematicians, Lienard and Chipart,[118] investigated the redundancy of the Routh–Hurwitz criteria and deduced that the necessary and sufficient conditions for the polynomial (eqn. 3.56) to have roots with negative real parts can be given in any one of the following forms:

(a) $a_n > 0, a_{n-2} > 0, \ldots ; \Delta_1 > 0, \Delta_3 > 0, \ldots$

(b) $a_n > 0, a_{n-2} > 0, \ldots ; \Delta_2 > 0, \Delta_4 > 0, \ldots$

(c) $a_n > 0, a_{n-1} > 0, a_{n-3} > 0 \ldots ; \Delta_1 > 0, \Delta_3 > 0, \ldots$

(d) $a_n > 0, a_{n-1} > 0, a_{n-3} > 0 \ldots ; \Delta_2 > 0, \Delta_4 > 0, \ldots$

These conditions have the advantage of requiring only the evaluation of about half the number of determinental inequalities as do the Hurwitz conditions.[119] They still, however, involve n inequalities. Proofs of Lienard's and Chipart's conditions have been given by Gantmacher[120] and Fuller.[121]

Unknown to Routh and Hurwitz, Russian mathematicians and engineers were also working on the problem of dynamic stability. The work had been started by Wischnegradski at the Practical Technological Institute in St. Petersburg in the 1870s with his analysis of the Watt governor; by the mid-1880s N. E. Zukowskii at the University of Moscow was also lecturing on the theory of control system. The major breakthrough, however, came in 1892 when Alexandr Michailovich Liapunov (1857–1918), a former student of P. L. Chebyshev at the University of St. Petersburg and now Professor of Mechanics at the University of Khar'kov, presented his doctoral dissertation on '*The general problem of the stability of motion*'.[122]

Liapunov, in addition to being closely connected with the Russian work, was also aware of the work on dynamic stability being done outside Russia. He frequently cites the work of Routh, as well as works by Hermite, and in the introduction acknowledges his debt to Poincaré.[123] The importance of Liapunov's work is that the methods he developed are applicable to nonlinear systems. His work, however, remained largely unknown in the English-speaking world until after the Second World War.[124]

References and notes

1 Boulton to Watt, 28 May 1788, quoted from DICKINSON, H. W., and JENKINS, R.: *James Watt and the steam engine* (Oxford University Press, 1927), p. 220

2 WATT, J., appendix to ROBISON, J.: *A system of mechanical philosophy* (Edinburgh, 1822, 5 vols., vol. 2, p. 151

3 DALTON, J.: *Descriptive poem addressed to two ladies at their return from viewing the mines near Whitehaven*, 1755, quoted from ROLT, L. T. C.: *Thomas Newcomen: the prehistory of the steam engine* (David & Charles, Newton Abbott, 1963), p. 70

4 MAYR, O.: 'Adam Smith and the concept of the feedback system', *Technology & Culture*, 1971, **12**, pp. 1–22

5 SMITH, A.: *An inquiry into the nature and causes of the wealth of nations*, 1776 (Routledge, London, 1894), p. 44

6 John Robison to James Watt, 22 October 1783, quoted in ROBINSON, E., and MCKIE, D.: *Partners in science: letters of James Watt and Joseph Black* (Constable, London, 1970), p. 30

7 *ibid.*, p. 249

8 FAREY, J.: *A treatise on the steam engine, historical, practical and descriptive* (Longmans, London, 1827, reprinted by David & Charles, Newton Abbott, 1971), p. 465

9 *ibid.*, p. 466

10 AIRY, G. B.: 'On the regulator of the clock-work for effecting uniform movement of equatoreals', *Mem. Roy. Astron. Soc.*, 1840, **11**, pp. 249–267

11 FAREY: *op. cit.*, p. 466

12 YOUNG, T.: *Lectures on natural philosophy and the mechanic arts* (Johnson, London, 1807, 2 vols.), vol. 1, p. 47

13 PONCELET, J. V.: *Course de mécanique appliquée aux machines* Kretz, M. X., (Ed.) (Gauthier-Villars, Paris, 1874 – it was originally issued in lithograph form in 1826, 1832, and 1836

14 This term seems to have been used by French writers; see footnote by Kretz in Poncelet, *op. cit.*, p. 100 and DWELSHAUVERS-DERY, V. A. E.: 'A new method of investigation applied to the action of steam engine governors', *Proc. ICE*, 1888, **94**, p. 212

15 The German term used is 'Ungleichförmigkeitsgrad', now translated as 'governor regulation'; RÖRENTROP, K.: *Entwicklung der modernen Regelungstechnik* (Oldenburg, München, 1971), pp. 33–34

16 HARTNELL, W.: 'On governing engines by regulating the expansion', *Proc. I. Mech. E* 1882, p. 414

17 MAYR, O.: 'Victorian physicists and speed regulation: an encounter between science and technology', *Notes and Records of the Royal Society of London*, 1971, **26**, p. 212

18 FULLER, A. T.: 'The early development of control theory', *J. Dynamic Systems, Measurement, and Control, Trans. ASME*, 1976, **98**, pp. 109–118; BENNETT, S.: 'A note on the early development of control theory', *J. Dynamic Systems, Measurement and Control, Trans. ASME*, 1977, **99**, pp. 211–213

19 Fraunhofer's work was widely reported; see FULLER: *Control theory – I*, p. 112

20 AIRY: *op. cit.* p. 250

21 FULLER: *Control theory – I*, deals with this paper in detail, and this account is largely drawn from Fuller

22 Quoted from FULLER: *Control theory – I*, p. 113

23 *ibid.*, p. 28

24 FULLER, A. T.: 'The early development of control theory – II', *J. Dynamic Systems, Measurement, and Control, Trans. ASME*, series G, 1976, **98**, p. 226; Mayr; 'Victorian physicists', p. 207

25 AIRY, G. B.: Supplement to paper 'On the regulation of the clock-work for effecting uniform movement of equatoreals', *Mem. Roy. Astron. Soc.*, 1851, **20**, pp. 115–119

26 *ibid*, p. 116

27 *ibid*, p. 117

28 See MAYR: 'Victorian physicists', pp. 215–217 and FULLER: *Control theory – II*, p. 228

29 THOMSON, W.: 'On electrically impelled and electrically controlled clocks', *Proceedings of the Glasgow Philosophical Society*, 1868, 6, pp. 61–64, paper read 24 January 1866

30 THOMSON, W.: 'On a new astronomical clock with free pendulum and independently governed motion for escapement wheel', *Proc. Roy. Soc.*, 1869, **17**, pp. 468–470; *Philosophical Magazine*, 1869, **38**, pp. 393–395

31 THOMSON, W.: 'On a new form of astronomical clock with free pendulum and independently governed uniform motion for escapement wheel' *in* Report of 46th Meeting of British Association for the Advancement of Science Glasgow 1876, (London, 1877). Reprinted in *Nature*, 1876–1877, **15**

32 British Patents 1870/3069 and 1871/810

33 THOMSON, W.: 'On a new form of centrifugal governor', *Transactions of Institution of Engineers in Scotland*, 1868, **12**, pp. 67–69; *Engineering*, 1869, 7, pp. 1–2

34 MAYR: 'Victorian physicists', p. 216

35 MAXWELL, J. C.: *On the stability of motion of Saturn's rings* (Macmillan, Cambridge, 1859). Reprinted in MAXWELL, J. C.: *The scientific papers of James Clerk Maxwell* (Cambridge University Press, 1890, and Dover, New York, 1965)

36 *ibid.*, pp. 295–296

37 *ibid.*, p. 296

38 THOMSON, W., and TAIT, P. G.: *Treatise on natural philosophy* (Oxford University Press, 1867)

39 MAYR, O.: 'Maxwell and the origins of cybernetics', *ISIS*, 1971, **62**, p. 429

40 THOMSON and TAIT: *op. cit.*, p. 280

41 MAYR: 'Maxwell', p. 429. Maxwell from 1865 until 1871 lived in semi-retirement, at Glenair, his country seat near Glasgow, where he worked mainly on his book, a *Treatise on electricity and magnetism*. He was, however, in the habit of making an annual visit to London, and, also, being close to Glasgow, kept in touch with William Thomson

42 *Proceedings of the London Mathematical Society*, 1868, **2**, pp. 60–61

43 *ibid.*, p. 61

44 MAXWELL, J. C.: 'On governors', *Proc. Roy. Soc.* 1867/8, **16**, pp. 270–283. The paper also appeared in *Philosophical Magazine*, 4th series, 1863, **35**, pp. 385–398; in Maxwell's *Scientific Papers*, NIVEN, W. D. (Ed.) (Cambridge Univerity Press, 1890, 2 vols.), vol. 2, pp. 105–120, and more recently in BELLMAN, R., and KALABA, R. (Eds.). *Mathematical trends in control theory*, (Dover, New York, 1964), pp. 5–17

45 Maxwell refers to the paper by SIEMENS, C. W.: 'On uniform rotation', *Philos. Trans. Roy. Soc.*, 1866, **156**, pp. 657–670, pls. 29 and 30

46 MAXWELL: *op. cit.*, 1868, p. 271 (page references are to the *Proceedings of the Royal Society*)
47 *ibid.*, p. 272
48 Maxwell's paper has been discussed in detail by MAYR: 'Maxwell', pp. 425–555; FULLER: *Control theory – II*, pp. 224–235
49 MAXWELL: *op. cit.*, 1868, p. 276
50 FULLER: *Control theory – II*, p. 233
51 MAXWELL: *op. cit.*, 1868, p. 278
52 FULLER: *Control theory – II*, p. 232, has suggested a possible arrangement
53 MAXWELL: *op. cit.*, 1868, p. 279
54 FULLER: *Control theory – II*, p. 229
55 REULEAUX, F.: 'Zur Regulatorfrage', *Zeitschrift des Vereins deutscher Ingenieure*, 1859, **3**, pp. 165–168. Realeaux is best known to English speakers through his book *Theoretische Kinematik: Grundzüge einer Theorie des Maschinenwesens* published in Germany in 1875, translated into English by A. B. W. Kennedy and published in 1876 as *The kinematics of machinery: outlines of a theory of machines*, (Macmillan, London, reprinted by Dover, New York, 1963)
56 LUDERS, J.: 'Über die Regulatoren', *Zeitschrift des Vereins deutscher Ingenieure*, 1861, **5**, pp. 60–75
57 LUDERS, J.: 'Über die Regulatoren', *Zeitschrift des Vereins deutscher Ingenieure*, 1865, **9**, pp. 401–416
58 KARGL, L.: 'Zur Lösung der Regulatorfrage', *Civilingenieur*, 1871, **17**, Pt. 1, pp. 265–296, Pt. 2, pp. 386–409, Plates 17 and 26
59 KARGL, L.: 'Beweis der Unbrauchbarkeit sämtlicher astatischer Regulatoren', *Civilingenieur*, 1873, **19**, pp. 421–430
60 Other transliterations are Wischnegradsky and Vyschnegradskiy
61 WISCHNEGRADSKI, J.: 'Mémoire sur la théorie générale des régulateurs', contributed to Académie des Sciences by M. Tresca, 1876. 'On direct-action regulators' (In Russian), *News of the Petersburg Institute of Practical Technology*, 1877; 'Über direktwirkende Regulatoren', *Civilingenieur*, 1877, **28**, pp. 96–132; 'Mémoire sur la théorie générale des régulateurs', *Revue Universelle des Mines*, Pt. 1, 1878, **4**, pp. 1–38, Pt. 2, 1879, **5**, pp. 192–227
62 A summary is given in PONTRYAGIN, L.: *Ordinary differential equations* (Adison-Wesley, New York, 1962), pp. 213–220, but this is very much modernised in approach
63 WISCHNEGRADSKI: *op. cit.*, 1878, pp. 12–13. 'Equation [3.20] which gives u as a function of time t, is a third order linear equation with constant coefficients, its solution depends, as is well known, on the roots of the equation $\theta^3 + M\theta^2 + N\theta + KLg/I\omega_0 = 0$. To be brief we shall give this equation the name characteristic . . .'
64 RÖRENTROP: *op. cit.*, p. 101
65 WISCHNEGRADSKI: *op. cit.*, 1878, p. 26
66 *ibid.*, p. 28
67 Maxwell's paper 'On governors' was not widely known in German-speaking countries; see PROFOS, P.: 'Professor Stodola's contribution to control theory', *J. Dynamic Systems, Measurement, and control, ASME*, series G, 1976, **98**, p. 119, and Routh's book *Advanced rigid dynamics* was not translated into German until 1898, as *Die Dynamik der Systeme starrer Körper*, (A. Schepp, Leipzig)

68 For further details of continental developments, see RÖRENTROP: *op. cit.*

69 For biographical details and a list of published work, see FULLER, A. T.: 'Edward John Routh', *Int. J. Control*, 1977, **26**, pp. 169–173

70 For a discussion of the close contacts between Maxwell, Thomson, Airy etc., see MAYR: 'Victorian physicists'

71 ROUTH, E. J.: 'Stability of a dynamical system with two independent motions', *Proceedings of the London Mathematical Society*, 1874, **5**, pp. 92–99. Reprinted in FULLER, A. T. (Ed.): *Stability of motion* (Taylor & Francis, London, 1976), pp. 141–143

72 ROUTH, E. J.: *A treatise on the stability of a given state of motion* (Macmillan, London, 1877. Reprinted in Fuller: *Stability of motion*)

73 *ibid.*, p. 12 (p. 42 in reprint)

74 *ibid.*, p. 24 (p. 54 in reprint)

75 See FULLER: *Stability of motion*, Introduction, pp. 1–18. Fuller gives proof of the special case and suggests that Routh probably obtained the theorem from a paper by Sturm (which is reprinted in the book)

76 ROUTH: *op. cit.*, p. 25

77 *ibid.*, p. 26

78 FULLER: *Stability of motion*, p. 13

79 Fourier had lectured on this topic long before he published his results; see TODHUNTER, I.: *An elementary treatise on the theory of equations* (Macmillan, London, 1875, 3rd edn.) p. 130

80 FULLER: *Stability of motion*, p. 10

81 STURM, C.: 'Other demonstrations of the same theorem', *J. de Mathématiques pures et appliquées*, 1836, **1**. Quotation is from translation by Fuller: *Stability of motion*, p. 204

82 ROUTH: *op. cit.*, p. 26

83 *ibid.*, p. 27 (p. 57 in reprint); this is Routh's first rule, see e.g. BROWN, B. M.: *The mathematical theory of linear systems* (Chapman & Hall, London, 1965, 2nd edn.) p. 97

84 *ibid.*, p. 29 (p. 59 in reprint)

85 *ibid.*, p. 34 (p. 64 in reprint)

86 ROUTH, E. J.: *An elementary treatise on the dynamics of a system of rigid bodies*, (Macmillan, Cambridge, 1860, 1st edn.; Macmillan, London, 1868, 2nd. edn., 1877, 3rd edn.). For the 4th edition the book was divided into two volumes, the whole being known as *A treatise on the dynamics of a system of rigid bodies*. Editions of the elementary part (Part 1) were produced in 1882 (4th edn.), 1891 (5th edn.), 1897 (6th edn.), 1905 (7th edn.) and for the advanced part (Part 2) in 1884 (4th edn.), 1892 (5th edn.), 1905 (6th edn.)

87 BOURNE, J.: *The steam engine* (Artisan Club, London, 1846; Longmans, London, 1861, 2nd edn.), passage quoted from Fuller: *Control theory – II*, p. 225. The work referred to was TREDGOLD, T.: *The steam engine* (Taylor, London, 1827)

88 *The Engineer*, 1859, **8**, pp. 119, 171; 1862, **14**, pp. 11, 307, 347, 363; 1866, **21**, p. 222; 1866, **22**, p. 147

89 ROUTH: *Rigid dynamics*, 3rd edn. 1877, p. 378; the sense of the passage is unchanged in the 5th edn. Part 2, 1892

90 *ibid.*, p. 379, and, in 5th edn., p. 76

91 SIEMENS, C. W.: discussion of paper by HEAD, J.: 'On the simple

construction of steam-engine governor having a close approximation to perfect action', *Proc. I Mech. E*, 1871, p. 226

92 SIEMENS, C. W.: discussion of paper by KITSON, F. W.: 'On the Allen governor and throttle valve for steam engines', *Proc. I Mech. E*, 1873, p. 62

93 ROUTH, E. J.: *Rigid dynamics*, p. 329 (3rd edn.), p. 76, (5th edn.), Part 2

94 HARTNELL, W.: 'On governing engines by regulating the expansion', *Proc. I Mech. E*, 1882, pp. 408–439, pp. 75–81

95 Note that the term is a misnomer since the quantity defined is 'work', not rate of work; later writers tended to use 'governor capacity'

96 RÖRENTROP: *op. cit.*, pp. 53–55; TRINKS, W.: *Governors and the governing of prime movers* (Van Nostrand, New York, 1919), p. 25

97 HARTNELL: *op. cit.*, p. 410

98 This account of Stodola's work is largely drawn from Profos: *op. cit.*, pp. 119–120

99 STODOLA, A. B.: 'Ueber die Regulierung von Turbinen', *Schweizer Bauzeitung*, 1893, **22**, pp. 113–117, 121–122, 126–128, 134–135; 1894, **23**, pp. 108–112, 115–117

100 HURWITZ, A.: 'Über die Bedingungen, unter welchen eine Gleichung nur Wurzeln mit negativen reelen Teilen bestizt', *Mathematische Annalen*, 1895, **46**, pp. 273–280; English translation by Bergmann, H. G.: 'On the conditions under which an equation has only roots with negative real parts' *in* BELLMAN, R., and KALABA, R.: *Selected papers on mathematical trends in control theory* (Dover, New York, 1964), pp. 72–82

101 HURWITZ: *op. cit.* (English translation), p. 75. The theorem by Hermite was published in 1856 – HERMITE, C.: 'Sur le nombre des racines d'une équation algébrique comprises entre des limites données', *J. reine angew. Math.*, 1856, **52**, pp. 39–51

102 HURWITZ: *op. cit.* (English translation), p. 73

103 BATEMAN, H.: 'The control of an elastic fluid', *Bull. Am. Math. Soc.*, 1945, **51**, p. 612. Reprinted in BELLMAN, R., and KALABA, R.: *op. cit.*, p. 30

104 HURWITZ: *op. cit.* (English translation), p. 72

105 STODOLA, A. B.: 'Erwiderung: Zur Frage du Regulierung hydraulischer Motoren', *Schweitzer Bauzeitung*, 1894, **23**, pp. 55–55, 71, 82, 89.

106 STODOLA, A. B.: 'Die amerikanischen Inertie-Regulatoren', *Schweizer Bauzeitung*, 1899, **28**, p. 178, and 'Des Siemenssche Regulierprinzip und die amerikanischer Inertie-Regulatoren', *Zeitschrift des Vereins deutscher Ingenieure*, 1899, **43**, Pt. 1, pp. 506–516, 573–579

107 PROFOS: *op. cit.*, p. 120

108 TOLLE, M.: *Die Regelung der Kraftmaschinen* (Springer, Berlin, 1905; 2nd edn., 1909, 3rd edn. 1922)

109 Trinks: *op. cit.*

110 *ibid.*, p. 134

111 SMITH, R. B.: discussion, CAUGHEY, R. J.: 'Steam-turbine governors', *Trans. ASME*, 1940, **62**, p. 197; VON FREUDENREICH, J.: 'Untersuchung der stabilitat von Regelvorrichtungen', *Stodola Festschrift* (Orell Fussli, Zurich, 1929), pp. 172–179; STEIN, Th.: 'Selbstreglung ein neues Gesetz der Regeltechnik', *Zeitschrift des Vereins deutscher Ingenieure*, 1928, **72**, pp. 165–171; SCHWENDNER, A. F., and LUOMA, A. A.: 'Superposed turbine regulation problem', *Trans. ASME*, 1936, **58**, paper FSP 58-8, pp. 6-5-6–20

112 HIGGS-WALKER, G. W.: 'Some problems connected with steam turbine governing', *Proc. I Mech. E*, 1941, **146**, pp. 117–125

113 *ibid*, p. 121

114 BRYAN, G. H., and WILLIAMS, W. E.: 'The longitudinal stability of aeroplane gliders', *Proc. Roy. Soc.*, 1904, **73**, pp. 100–116, paper read 18 June 1903. The reference to Routh was to *Advanced rigid dynamics* (edition not given)

115 HOPKINSON, B.: 'The "hunting" of alternating-current machinery', *Proc. Roy. Soc.*, 1904, **72**, pp. 235–252

116 BOMPIANI, E.: 'Sulle condizioni sotto le quali un equazione a coefficienti reale ammette solo radici con parte reale negative', *Giornale di Matematica*, 1911, **49**, pp. 33–39; see BATEMAN: *op. cit.*, p. 612. FRAZER, R. A., and DUNCAN, W. J., in a paper 'On the criteria for the stability of motion', *Proc. Roy. Soc.*, 1929, **124**, pp. 642–654, developed from the work of Routh criteria equivalent to the Hurwitz criteria with the determinants changed by transposing rows and columns

117 ORLANDO, L.: 'Sul problema di Hurwitz', *Rendiconti Accademia Lincei*, 1910, **19**, pp. 801–803; *Mathematische Annalen*, 1911, **71**, pp. 233–245; see BATEMAN: *op. cit.*, p. 612

118 LIENARD, A., and CHIPART, H.: 'Sur la signe de la partie réele des racines d'une équation algebrique', *J. Math. Pures Appl.*, 1914, **10**, pp. 291–346

119 GANTMACHER, F. R.: *The theory of matrices* (Chelsea, New York, 1959, 2 vols; published in Russian in 1954), vol. 2, p. 221

120 *ibid.*, pp. 220–225

121 FULLER, A. T.: 'Stability criteria for linear systems and realizability criteria for *RC* networks', *Proceedings of the Cambridge Philosophical Society*, 1957, **53**, pp. 878–896. A further investigation of redundancy in stability criteria has recently been made by FULLER, A. T.: 'On redundance in stability criteria', *Int. J. Control*, 1977, **26**, pp. 207–224. For general accounts of stability criteria, see GANTMACHER: *op. cit.*, pp. 172–250; JURY, E. I.: *Inners and stability of dynamic systems* (Wiley, New York, 1974)

122 The dissertation was published in 1892 by the Khar'kov mathematical society. It has not been translated into English, but in 1907 a French translation entitled 'Le problème général de la stabilité du mouvement', *Annales de la Faculté des Sciences de Toulouse*, 1907, **9**, which incorporated a number of corrections and changes was produced. This French version was photolithographically reissued in 1947 as *Mathematical studies No. 17* (Princeton University Press)

123 SILJAK, D. D.: 'Alexandr Michailovich Liapunov (1857–1918)', *J. of Dynamic Systems, Measurement, and Control, Trans. ASME*, 1976, **98**, p. 121

124 Minorsky makes reference to the French translation of Liapunov's dissertation in a paper published in 1941, MINORSKY, N.: 'Control problems', *J. Franklin Inst.*, 1941, **232**, pp. 543–545

Chapter 4

The development of servomechanisms

> It is nearly as hard for the practitioners of the servo art to agree on a definition of a servo as it is for a group of theologians to agree on sin.

I. A. Gettings, 1945

Introduction

The high festival of Victorian England, the Great Exhibition of 1851, demonstrated, so it was thought, Britain's position as the leading nation of the world. It was a celebration of the powers of steam; the exhibits were the products of the factory system, that 'vast automaton, composed of various mechanical and intellectual organs, acting in uninterrupted concert for the production of a common object all of them subordinate to a self-regulated moving force [the steam engine]'.[1] The handbills were printed on steam presses, the visitors were conveyed by steam locomotive – for, by 1851, the railway served every major town in England but Hereford, Yeovil and Weymouth – or were brought across the seas by steamship. 'If England today were to be deprived of her steam engines, she would also be robbed of coal and iron; all her sources of wealth would be cut off and all her means of development would be destroyed; this would mean the ruin of the vast Power. The destruction of her fleet, which is regarded as her most certain protection, would perhaps be less fatal for England', wrote Sadi Carnot in 1824.[2] By 1851 the fleet itself was becoming dependent on steam.

The era of the large steamship began in October 1835,[3] when, at a meeting of the Great Western Railway Company, in response to a director who expressed misgivings at the length of the proposed main line from London to Bristol, the engineer to the company, Isambard Kingdom Brunel, retorted: 'Why not make it longer and have a steamboat go from Bristol to New York and call it the Great Western'.[4] Two years later the *Great Western*, displacing 1340 tons and with engines of 750 h.p., the largest and most powerful ship yet built, was launched. The *Great Britain* followed in 1845 and in 1859 the *Great Eastern*, 21 000 tons and with engines of 8300 h.p. (6·2 MW) was launched.

The *Great Eastern*'s screw engines were originally to have been regulated by the marine-engine governor which Brunel's father had patented in 1822. 'I remember the experimental governor being made and my working at it, I made the drawings for a patent', Isambard wrote in a letter to the Patent Office.[5] The ship was also to have had a stabilised platform for astronomical observations – the platform was to have been mounted on gimbals and stabilised by means of a flywheel revolving at high speed. Brunel worked on this idea from 1852 to 1854, corresponding extensively with the Astronomer Royal, Sir George Airy, and Professor Piazzi, of the Edinburgh Observatory,[6] but increasing problems in the construction of the ship itself led to the abandoning of the proposal.

Some years later, in 1874, Henry Bessemer (1813–1898) made an ambitious attempt to stabilise the whole saloon of a ship. He used a gyroscope to sense the motion; power to move the saloon was obtained from the ship's hydraulic system.[7] A similar scheme, designed to provide a steady platform for the firing of guns, was devised by Beauchamp Tower in 1889.[8] In 1892 Sir John Thornycroft[9] (1843–1928) attempted to stabilise a whole ship. He used a pendulum to sense the motion of the ship, the movement of the pendulum controlling hydraulic machinery, which shifted a weight equivalent to 5% of the ship's displacement from one side of the hold to the other. However, successful ship stabilisation was not achieved until the 20th century; the major servomechanism developments during the 19th century were made in connection with the steering of ships.

Steering engines and position servomechanisms

I believe that a ship may be so stowed, and so managed, that she may be steered by the little finger, as I know a line-of-battle ship could have been. I cannot see why, in the present day, with all our science, we cannot secure the action of the ship herself before we act upon the rudder, so that she shall obey the h 'm literally as our former sailing ships would do.

Admiral Sir Edward Belcher, KCB

> A vessel under sail and properly trimmed is on a beautifully
> delicate edged balance; and no doubt some vessels are so beauti-
> fully trimmed that they can be put about either without the
> helm or with scarcely a touch on it. That is a condition which
> we do not meet with in steamers. They are not balanced at all. The
> whole of the work of turning them has to be done by the helm.

> Charles W. Merrifield, FRS
> [Contributions to discussion at a meeting of the Institution of
> Naval Architects, 1873]

In good conditions, sailing ships could be steered with little or no use
of the rudder; in heavy seas this ideal could not be met. For a large
ship the forces on the rudder – friction, inertia and, in a moving ship,
hydrodynamic forces – are considerable. Traditionally the helm was
operated manually and, owing to the forces involved, large gear ratios
between the helm and rudder had to be used; as a consequence, steering
was neither precise nor responsive. The lack of precision and responsive-
ness could not be tolerated in warships, but the use of lower gear ratios
to improve the 'feel' meant that 'in some of the armour plated ships
in the British Navy it requires nearly a hundred men to put the helm
hard over when the vessel is going at full speed'.[10] The solution appeared
to lie in the use of steam power to operate the rudder.

Steam steering engines

The first steam steering engine was invented by Frederick E. Sickels in
1849 and patented by him in 1853 and in 1860. After being exhibited
at an exhibition in New York in 1853 it was installed in a coastal
steamer, the *Augusta*, and used successfully for two years. The steering
engine was demonstrated in several other vessels before being exhibited
at the World's Fair in London in 1862.[11] Sickels's engine was, 'a crude
step-by-step servomechanism' without feedback.[12] •

A steering engine incorporating feedback was patented in 1866 by
J. McFarlane Gray[13] and, appropriately enough, first saw service in the
largest and most advanced ship then afloat, Brunel's *Great Eastern*.
Many years later Gray said of his invention that:

> The principal thing that I did was to make an automatic con-
> trolling valve, continuous in its action. To put a handle to where
> you wanted something to move to had been done before, but I
> saw that that would not do for steering. I therefore contrived the
> differential movement of the reversing valve, but [by?] which the
> rudder or other object to be moved would be made to follow the
> movement of the controlling wheel or lever.[14]

The arrangement of Gray's engine is shown in Fig. 4.1. The
movement of the helm is connected by a bevel gear to the input shaft

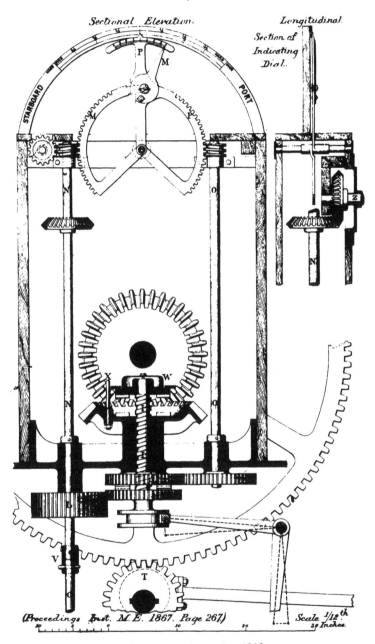

Fig. 4.1   *McFarlane Gray's steam steering engine, 1868*
[Reprinted by permission of the Council of the Institution of Mechanical Engineers from *Proc. I Mech. E*]

of the engine, rotation of the input shaft turns a pointer (P) to indicate the desired rudder angle and, because the rudder and hence the bevel gear are instantaneously fixed, it causes the differential screw to move linearly, thus opening the steam valve. If the input is now held stationary, the movement of the rudder, transmitted through the crown wheel and bevel gear, moves the differential screw so as to close the steam valve; the rotation of the bevel wheel is also transmitted by the feedback shaft to the pointer, which shows the angle of the rudder. As the block diagram (Fig. 4.2) shows, the system is a true servo-mechanism.

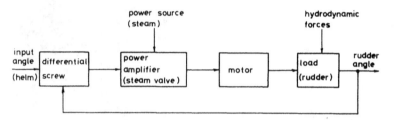

Fig. 4.2   *Block diagram of steam steering engine*

The word 'servomechanism' derives from the work of Joseph Farcot, who may have equal claim with Gray as the inventor of steering engines incorporating feedback. Jean Joseph Léon Farcot, a member of the firm of Farcot and Son, worked with his father, Marie Joseph Denis Farcot, on governors for steam engines. In 1854 his father patented two methods of eliminating offset by modifying Watt's governor, and father and son jointly obtained further governor patents in 1862 and 1864, the latter for a spring-loaded governor.* In 1868 Jean Joseph obtained a British patent for a steering engine, and in 1873 he published a book *Le servo-moteur ou moteur asservi*, in which he describes the various designs of steam steering apparatus developed by the company of Farcot and Son.[15]

Farcot described the particular principle of his invention in the following way [the italics are his]:

> Asservir tout moteur au gouvernement absolu d'un conducteur *en faisant cheminer* directment au par un intermédiaire quelconque, *la main de celui-ci, avec l'organe sur lequel agit le moteur*, de telle sorte que tous deux marchent, s'arrêtent, reculent, reviennent ensemble, et que le moteur suive pas à pas le doigt indicateur du conducteur dont il imite servilement ou les

---

* See chapter 2.

gestes. Nous avons cru qu'il était nécessaire de donner un nom nouveau et caractéristique à cet engin nouveau et l'avons appelé *servo-moteur* ou *moteur asservi*.[16]

The servomotor arose as a consequence of Farcot's attempts to devise governors with sufficient power to operate the valves of marine engines of 500 to 1000 h.p. (400 to 800 kW). He had attempted to devise a relay governor by interposing between the governor and the valve mechanism a steam cylinder; the governor operated a slide valve, which admitted steam to the cylinder, the piston of which was connected to the engine valve. This mechanism was totally unsatisfactory: 'devient tout à fait instable dans son mouvement et passe instantément, sans motif, ou pour la moindre ouverture du tiroir, d'une extremite à l'autre de sa course.'[17] It was in solving this problem that Farcot was led to the discovery of a new engine, a servomotor.

Farcot's basic servomotor is shown in Fig. 4.3. Lever *u* operates a slide valve, allowing steam to flow to the appropriate side of the piston; the piston is connected, by means of the slider crank, to shaft *j*, which in a steering engine would be connected to the rudder. The input to the device is by movement of lever *m*, which by means of a bell crank opens and closes the slide valve. If *m* is moved to the left, *u* moves to the right and the slide valve admits steam to the right-hand side of the piston; as the piston moves to the left, point *o*, which is attached to a flange on the arm moved by *v*, rotates until the arm so is once again vertical; in doing so it shuts off the slide valve.

The work of Joseph Farcot represents an important step in the development of control engineering, for not only were his inventions and designs of practical importance, but his book was the first extensive account of the general principles of position control mechanisms. He notes that the servomotors described can be used for a variety of tasks besides the steering of ships; an example he gives is for use in raising weights, as shown in Fig. 4.4. As well as his basic design he also gives examples of modifications; for example, a 2-stage system (Fig. 4.5) for use on larger engines and a system designed by one of his licensees, M. Duclos (Fig. 4.6), which uses direct feedback from the piston movement instead of a differential lever. In describing the steering engine installed in the coastal defence ship *Bélier* he notes that the helmsman is required to exert a force of 3 to 4 kg to operate the rudder, while the rudder loads can be as high as 10 000 to 12 000 kg. Farcot also recommends that the slide valve be made with an overlap of 1 to 2 mm, which, he says, corresponds to an 'angle d'indétermination' of ± 1°.[18]

By 1872 Farcot's steering engines had been fitted to five small

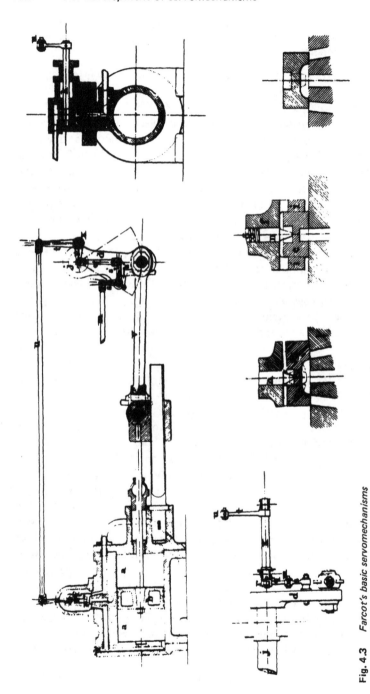

Fig. 4.3   *Farcot's basic servomechanisms*
[Farcot, J. J.: *Le servo-moteur ou moteur-asserir* (1873)]

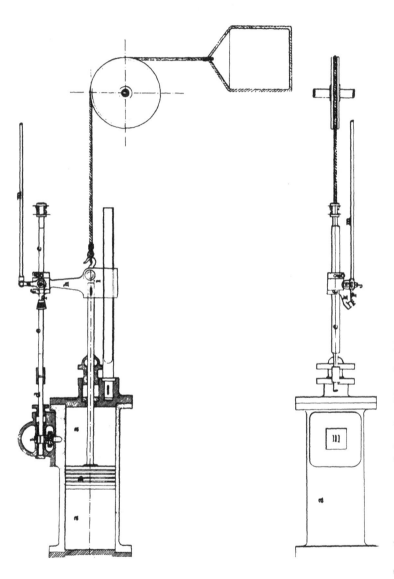

Fig. 4.4   *Farcot's servomotor applied to the raising of weights* [Farcot, *op. cit.*]

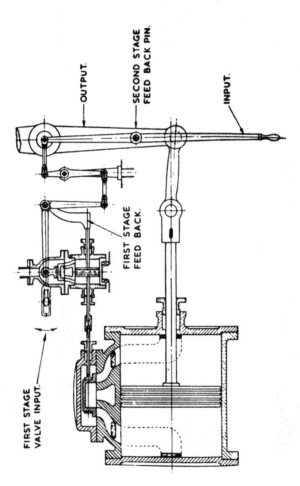

Fig. 4.5   *Farcot's 2-state servomechanism*
[Farcot, *op. cit.*]

ships of the French Navy; these ships were also equipped, by Farcot, with engines for rotating the gun turrets. These engines were possibly based on the rotary servomotor designed by Farcot and shown in Fig. 4.7. Farcot gives no details of this motor, presumably, it has been suggested, because of security regulations.[19]

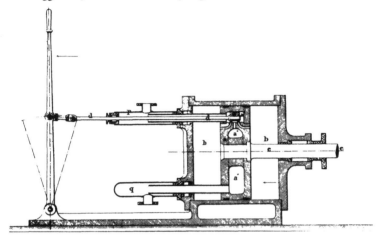

Fig. 4.6 *Duclos's improved servosystem*
[Farcot, *op. cit.*]

Across the channel the British Admiralty was also alert to the use of steam for the operation of gun turrets: H.M.S. *Thunderer* was so fitted in 1872. However, steam was not an entirely suitable medium, as *The Times* explained: 'The only defect is in the steam machinery which rotates the turrets, the elasticity of the motive power rendering it occasionally difficult to stop the guns precisely in the loading position . . . . In the *Inflexible*, however, the rotating as well as the loading gear will be worked by hydraulic power, a power which is completely under the command of the operator.'[21]

Farcot, in order to reduce steam consumption, had developed a mixed steam–hydraulic system for steering engines, shown in Fig. 4.8. The hydraulic part of the system acted as a locking mechanism, for, when the valve N was closed, the piston B was held fast by the hydraulic forces; to prevent damage from excessive forces on the rudder, the system was fitted with relief valves.

Hydraulic steering engines and gun turrets
In the late 1860s the British Admiralty equipped several of their sailing ships with various types of steering engines; one ship, H.M.S. *Minotaur*,

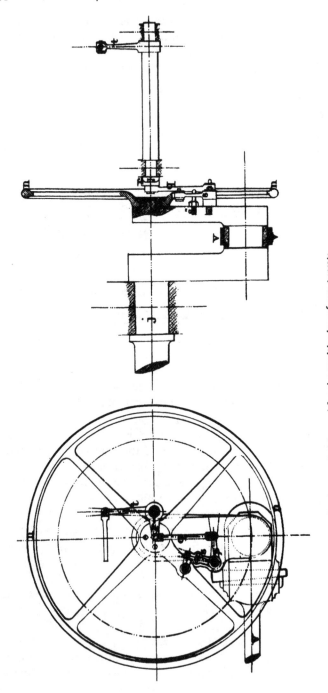

Fig. 4.7   Farcot's rotary servomechanism, probably used for the positioning of gun turrets
[Farcot, op. cit]

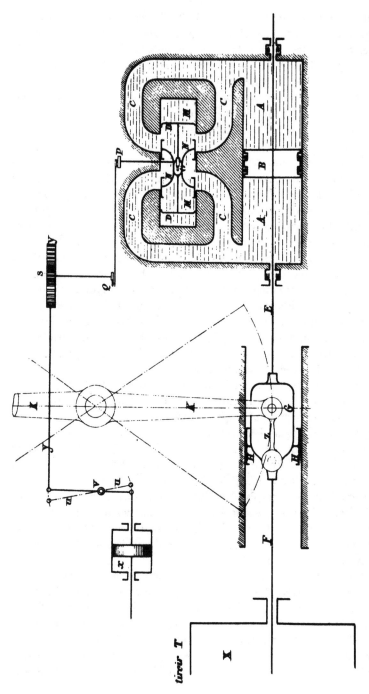

Fig. 4.8   *Farcot's steam-hydraulic servosystem*

had a steam–hydraulic system. A small steam engine was used to obtain the necessary head of water for the hydraulic rams used to operate the rudder. This led to an ingenious system designed by Captain E. A. Inglefield in 1869 and fitted to H.M.S. *Achilles*.[21] Inglefield's system made use of the static pressure due to the head of sea water obtainable at the bottom of the hull, the idea being to make the steering independent of the operation of the steam engine. Independence was, of course, only available for a limited period of time since, in order to use the static head of pressure, discharge of water into the bilge was necessary, and hence pumps were required to remove this water to allow continuous operation. His system included feedback, which he described in the following terms: 'An indicator placed before the helmsman shows the amount of movement of the helm, and the moment the tiller has been placed over to the desired angle the slide is made to lock the water into its position, and the helm is at once rigid'.[22]

In 1870, Andrew Betts Brown patented a hydraulic servomechanism for use in connection with the steering of ships. In his patent he discloses that the actuating mechanism is operated by 'hydraulic cylinders in conjunction with a valve moved by the [input] and controlled by the position motion of the [output]'.[23] Brown's mechanism is shown in Fig. 4.9, and in a patent of 1871[24] he described a 2-stage valve shown in Fig. 4.10. These hydraulic valves were developed for steering engines and hydraulic cranes for merchant ships.

During the same period G. W. Rendel (1833–1902) of Sir. W. G. Armstrong and Co. was developing hydraulic machinery for naval use – the Dutch gunboat *Hydra* was equipped with hydraulic machinery in 1874, and two years later similar equipment was installed in H.M.S. *Dreadnought*. For operating gun turrets Rendel designed a 3-cylinder hydraulic motor. Rotary piston motors were also designed by A. B. Brown, James Hastie and Peter Brotherhood.[25] Interest in hydraulic servomechanisms was also developing in Germany; in 1879 Lincke published a long article discussing the kinematics of relay systems in which he shows two forms of feedback relays, (Fig. 4.11). His concern was not with the details of the practical applications, but with the analysis of relay systems.[26]

These early hydraulic servomechanisms were constant-pressure systems with fixed-displacement pumps and accumulators. They had the disadvantage that the pump and accumulator had to be of sufficient size to supply the flow necessary during extensive manoeuvring; consequently, during long periods when there was little rudder movement, the pump bypass would be open. For this reason hydraulic steering engines do not seem to have been widely used; the United States

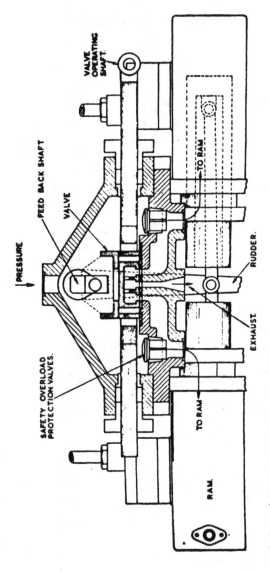

Fig. 4.9   *A. B. Brown's hydraulic steering engine, 1870*
[Reprinted from Conway, H. G., *Trans. Newcomen Soc.* 1953–1955, 29]

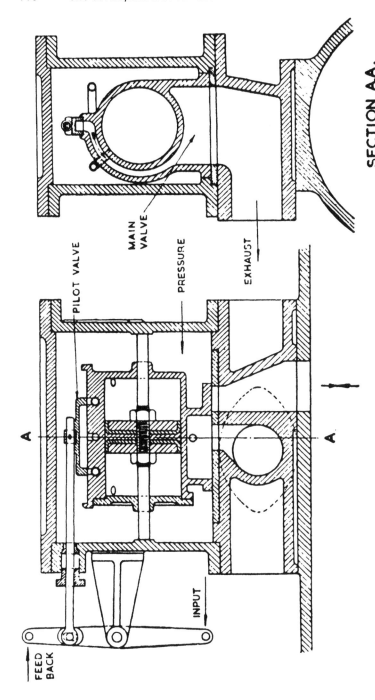

Fig. 4.10   A. B. Brown's 2-stage servomechanism
[Reprinted from Conway, H. G., Trans. Newcomen Soc., 1953–1955, **29**]

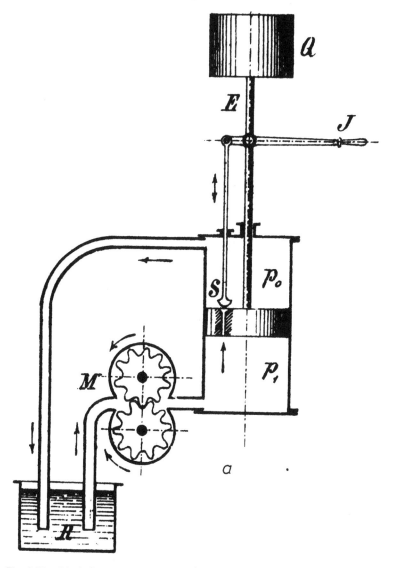

Fig. 4.11 *Lincke's servomechanisms, 1879 [part b on p. 112]*
[Reprinted from Lincke, F., *Zeitschrift des vereins deutscher Ingenieure,* 1879, **23**]

Navy continued to use steam steering engines, until the 1920s, although they experimented with electric motors.[27]

Hydrostatic transmission was introduced at the beginning of the 20th century with the development by the Waterbury Company of a

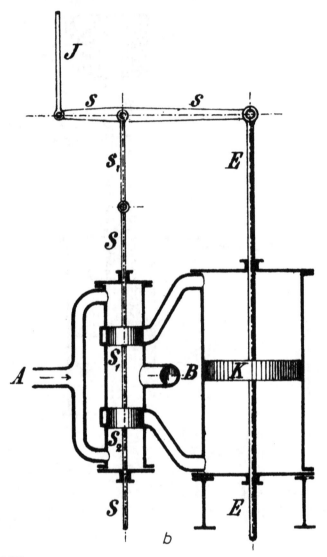

Fig. 4.11*b*

variable-delivery swash-plate pump based on a design by Williams and Janney. In 1906 a Waterbury transmission was used, in place of a Ward–Leonard motor generator set, to operate the elevation system of the 12-inch guns on the U.S.S. *Virginia*. In Britain, H. S. Hele-Shaw,

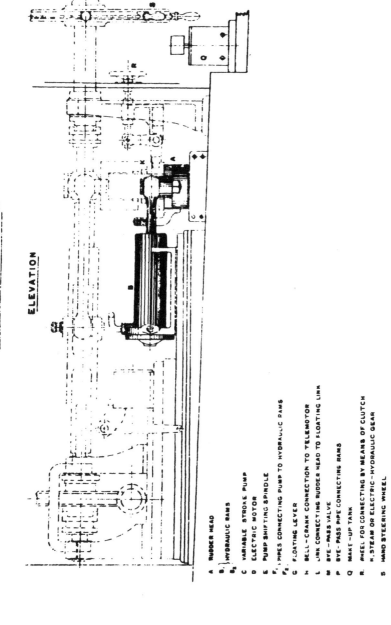

STEAM STEERING GEAR (DOTTED LINES) AND ELECTRIC-HYDRAULIC GEAR (FULL LINES)
OF TURBINE YACHT ALBION.

ELEVATION

A   RUDDER HEAD
B₁ HYDRAULIC RAMS
B₂
C   VARIABLE STROKE PUMP
D   ELECTRIC MOTOR
E   PUMP SHIFTING SPINDLE
F₁ PIPES CONNECTING PUMP TO HYDRAULIC RAMS
F₂
G   FLOATING LEVER
H   BELL-CRANK CONNECTION TO TELEMOTOR
L   LINK CONNECTING RUDDER HEAD TO FLOATING LINK
M   BYE-PASS VALVE
P   BYE-PASS PIPE CONNECTING RAMS
Q   MAKE-UP TANK
R   WHEEL FOR CONNECTING BY MEANS OF CLUTCH
H, STEAM OR ELECTRIC-HYDRAULIC GEAR
S   HAND STEERING WHEEL

Fig. 4.12   [See p. 114]

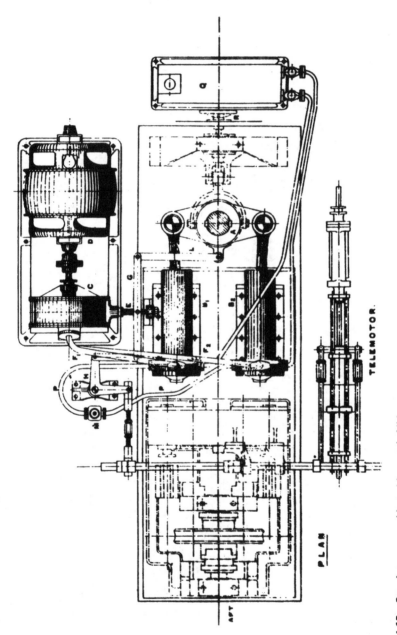

Fig. 4.12 *Steering gears used in turbine yatch Albion*
[Reprinted from Hele-Shaw, H. S., and Martineau, F., *Trans. Inst. Naval Arch.*, 1911, 53]

(1854–1941) and T. E. Beecham developed a variable-stroke radial piston pump which was first used in 1911 on a steering engine for the turbine-powered yacht *Albion*. The *Albion* had originally been fitted with an all-electric steering gear, but the continual stopping and starting of the motor had led to problems of rapid wear on the commutator of the dynamo, so the electrical system was replaced by a conventional steam steering engine.[28] Following Hele-Shaw's invention of a variable-stroke hydraulic pump, an experimental hydraulic steering system was installed in the *Albion* while retaining the existing steam-powered gear. Trials to compare the two systems were carried out in February 1911.

The arrangement of the hydraulic steering engine, which Hele-Shaw and Martineau refer to as an 'electric–hydraulic steering gear', is shown in Fig. 4.12. In order to compare the performance of the two systems,

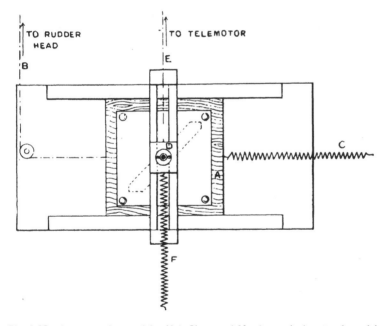

Fig. 4.13 *Lag recorder used by Hele-Shaw and Martineau during steering trials on the Albion*
[Reprinted from Hele-Shaw, H. S., and Martineau, F., *Trans. Inst. Naval Arch.*, 1911, **53**]

Hele-Shaw and Martineau needed to measure and record the relationship between the movement of the helm and of the rudder. Since they were not 'able to discover in the numerous contributions to the questions of steering, published in the Transactions of this Institution

[Naval Architects], any direct mode of measuring and recording this relation',[29] they were forced to devise their own recorder. They called it a lag recorder and it is shown in Fig. 4.13. The telemotor is a simple mechanical arrangement which converts the rotary motion of the helm to a linear movement which is used to operate the steam supply valve for the steam steering engine.

Using the recorder Hele-Shaw and Martineau obtained what are probably the first records ever made of the relationship between the input and output of a servomechanism. The diagrams given in their paper, a selection of the many which they recorded, are shown in Fig. 4.14

LAG DIAGRAMS

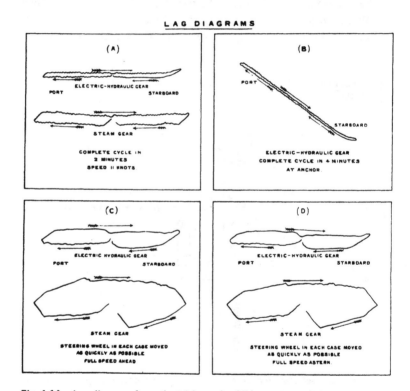

Fig. 4.14   *Lag diagrams from the trials on the Albion*
[Reprinted from Hele-Shaw, H. S., and Martineau, F., *Trans. Inst. Naval Arch.*, 1911, **53**]

(note that records A, C, and D have been rotated through 45°). They were obtained with what was in effect a triangular waveform input and they clearly show the difference in velocity lag between the hydraulic and steam steering engine.

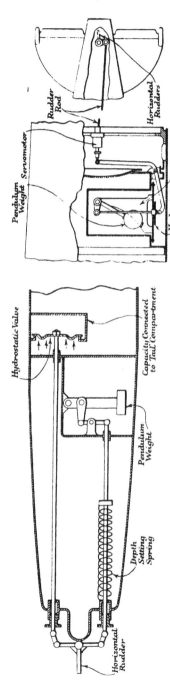

Fig. 4.15   [See p. 118]

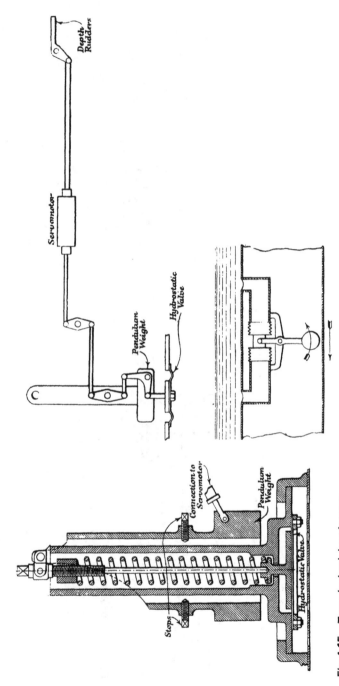

Fig. 4.15   *Torpedo depth-keeping gear*
[Reprinted from Bethell, P., *Engineering*, 9 Nov 1945, p. 365]

Torpedoes

The most advanced servomechanisms developed during the last quarter of the 19th century were those used in torpedoes. In 1864 Robert Whitehead (1823–1905), manager of Stabilimento Technico Fiumano, an engineering works at Fiume,* the Adriatic base for the Austrian Navy, was approached by a Captain Giovanni Luppis, who had devised a scheme for a self-propelled boat steered by long yoke lines from the shore. Whitehead, after constructing a model, considered the scheme to be impracticable and abandoned the idea. Three years later he demonstrated to the Austrian Government a weapon, propelled by a pneumatic engine, which ran at any chosen depth beneath the surface, was independent of the firing craft from the moment of launching and carried an explosive charge. Further demonstrations of the 'torpedo' were given at Fiume in 1869 and witnessed by British officers; in 1870 the Admiralty held an extensive series of trials and by 1872 Whitehead torpedoes were being built at Woolwich Arsenal.

The depth keeping of the torpedoes used in the 1867 trials was uncertain; Whitehead had used a hydrostatic valve to sense the depth of immersion, the movement of the valve being used to control the

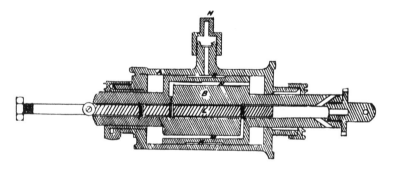

In the above diagram A = cylinder, B = piston, L = cylindrical slide valve.

Fig. 4.16   *Torpedo servomotor of about 1890*
    [Armstrong, G.: *Torpedoes and torpedo boats* (1901)]

horizontal rudders. By 1869 he had developed the system which, for 25 years, was referred to as 'the secret'; this was to make the setting of the horizontal rudders proportional not only to the depth of

* Now Rijeka-Susak.

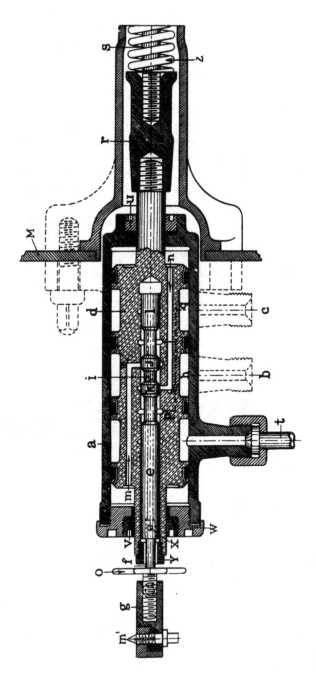

Fig. 4.17   *Torpedo servomotor of about 1900*
[Reprinted by permission of the Smithsonian Institution Press from *Feedback mechanisms*, O. Mayr, *Smithsonian studies in history and technology*: number 12: Figure 114. Washington, DC: Smithsonian Institution Press, 1971]

immersion but also to the attitude of the torpedo. A simple arrangement for combining the two signals is shown in Fig. 4.15. The attitude is sensed by the pendulum, and, since the rate of change of depth is proportional to the attitude, Whitehead had, in effect, introduced velocity feedback into the depth control system.

By 1876 the depth control system had taken on a modern appearance with the introduction of a pneumatic servoamplifier between the depth and attitude sensors and the horizontal fins;[30] an early version of this device (Fig. 4.16) was only four inches long, requiring half an ounce force to operate the slide valve and provided an actuating force of 180 lb.[31] Fig. 4.17 shows a much improved version which was fitted to the Whitehead torpedoes built in the USA.

Unlike the navies of the European countries, the US Navy made little investment in Whitehead torpedoes, largely because of the work of Commander John Adams Howell. Between 1870 and 1884 Howell developed a torpedo, which ran quietly and, unlike the Whitehead torpedo, did not leave a tell-tale trace of bubbles. The Howell torpedo obtained its power from a heavy flywheel which was spun up to a high speed (about 10 000 rev/min) before the torpedo was launched. It incorporated depth control apparatus similar to the Whitehead torpedo except that the pneumatic servoamplifier had to be replaced by a mechanical servoamplifier powered from the flywheel.

The Howell torpedo also incorporated a control system, for directional control; the heavy flywheel mounted rigidly in the centre of the torpedo acted as a constrained gyroscope and the horizontal forces acting to deflect the torpedo from its course therefore produced a couple tending to roll the torpedo about its horizontal axis. This roll was detected by means of a pendulum (Fig. 4.18) whose motion acting

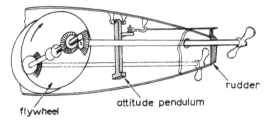

rudder

attitude pendulum

flywheel

Fig. 4.18 *Schematic showing flywheel drive and directional control mechanism of the Howell torpedo*

through a mechanical servoamplifier (not shown in Fig. 4.18), operated the rudder to correct for the disturbance. This appears to be the first successful use of the gyroscopic principle in a control system.

In 1895 Ludwig Obry of the Austrian Navy invented a gyroscopic device for use in torpedoes. The gyroscope was mounted with its axis of spin parallel to the longitudinal axis of the torpedo, as shown in Fig. 4.19. The relative movement between the vertical gimbal and

Fig. 4.19   *Obry's gyroscope for directional control*
[Reprinted from Bethell, P., *Engineering*, 19 Oct. 1945, p. 302]

the frame, owing to yaw, was used to operate, through a pneumatic relay, the vertical rudders. These early systems were on/off — the rudders were always hard over at one or other of the two stops and proportional action does not seem to have been introduced until the early part of the 20th century.[32] Directional gyroscopes of this type were fitted to Whitehead torpedoes in the USA in 1896 and, in 1897, the British Navy began using them in Whitehead torpedoes.[33]

Stabilisation and steering of ships

> With the gryoscope it is possible to create and maintain a very
> powerful fulcrum in space effective for the heaviest kind of
> mechanical duty.
>
> Elmer Sperry, 1910

Interest in the application of the gyroscope grew rapidly following
the publication of a series of articles in *Scientific American* in 1907.[34]
These articles reported on the work of the Germans, Schlick and
Anschütz-Kaempfe, and an Englishman, Brennan.

In 1904 Ernst Otto Schlick suggested placing a large gyroscope
in the hull of a boat in order to reduce rolling; trials of a gyrostabiliser
based on this idea took place aboard the *See-Baer*, a torpedo boat of
the German Navy, in 1906. Sir William White, sent by the British
Admiralty to observe the trials, reported so favourably that the firm of
Swan, Hunter and Wigham Richardson acquired the Schlick patents and
even purchased the *See-Baer* from the German government to use for
test purposes.[35] Schlick's gyrostabiliser was a passive device; it increased
the effective inertia of the ship by an amount proportional to $I^2\omega^2$,
where I is the inertia of the gyroscope flywheel and ω is its angular
velocity, and hence increased the period of natural oscillation of the
ship.

In 1903, following his participation in an Arctic expedition, Dr.
Hermann Franz Joseph Hubertus Maria Anschütz-Kaempfe (1872–1931),
a history-of-art graduate who had later studied medicine, made plans
for an expedition by submarine to the North Pole. Despite initial
scepticism his plans received some support when the Krupp shipbuilding
firm stated that it could build a suitable submarine. The outstanding
problem was that of the provision of a suitable guidance system.
Anschütz-Kaempfe attempted to use the tendency of a gyroscope freely
suspended at its centre of gravity to maintain a fixed position in space
as a basis for a guidance system. The first trials of this device made in
1903, although disappointing, showed sufficient promise to interest the
Germany Navy, and further tests followed in 1904. Encouraged by the
Navy, Anschütz-Kaempfe formed the Anschütz company in 1906, and,
with the help of a group of engineers and scientists, developed a gyrocompass. The excellent performance of the compass during sea trials
aboard the *Deutschland* in 1908 attracted the attention of naval engineers throughout the world.[36] One such engineer was Elmer Sperry.

Sperry had become interested in the gyroscope in 1907. He tried
to interest circus owners in the use of gyroscopes to enable acrobats to

perform spectacular feats and he also patented its use for steadying vehicles,[37] a proposal which was revived in the 1930s in connection with stabilising tanks. He was also fascinated by the gyrostabilised monorail which was demonstrated by Louis Brennan to the Royal Society in London, and he visited Brennan in 1909.[38]

In 1908 Sperry turned his attention to the use of the gyroscope on board ship; in May of that year he filed what proved to be the basic patent for the 'active stabiliser'.[39] Evidence from his notebook suggests that he was also working on a gyrocompass,[40] but work on the gyro-stabiliser delayed the compass and it was not patented until 1911. In England, S. G. Brown, assisted by John Perry, was working along similar lines, but it was not until 1916 that the Brown gyrocompass was patented.[41]

Stabilisers

In his patent application concerning the stabilisation of ships, Sperry wrote [my italics]

> It has been known ... that gyroscopes mounted in swinging frames in a plane transverse of a ship would reduce the rolling to some extent ... such apparatus was sluggish, did not act effectively until the ship had acquired considerable momentum and hence was only partially successful in dampening the rolling .... My invention becomes *active* promptly on incipient rolling.[42]

Sperry always referred to his system as 'active instead of passive' in distinguishing it from Schlick's. The arrangement of the 'passive' gyro-stabiliser is shown in Fig. 4.20*a*. The action of the ship in rolling about the *x*-axis produces a torque about the *z*-axis, given by

$$T_z = J_z \frac{d^2\theta_z}{dt^2} + B_z \frac{d\theta_z}{dt} - J_s\omega_s \frac{d\theta_x}{dt} \qquad (4.1)$$

where the direction is as shown in Fig. 4.20*b*. If the torque resisting the precession about the *z*-axis is $K_z\theta_z$, eqn. 4.1 becomes

$$J_z \frac{d^2\theta_z}{dt^2} + B_z \frac{d\theta_z}{dt} + K\theta_z = J_s\omega_s \frac{d\theta_x}{dt} \qquad (4.2)$$

As a result of precession about the *z*-axis, a torque T_x,

$$T_x = J_x \frac{d^2\theta_x}{dt^2} + B_x \frac{d\theta_x}{dt} + J_s\omega_s \frac{d\theta_z}{dt} \qquad (4.3)$$

occurs in the direction as shown in Fig. 4.20*c*.

The equation of the ship in rolling is given by

$$I\ddot{\theta}_x + B_s\dot{\theta}_x + WH\theta_x + T_x = F(t) \tag{4.4}$$

If J_z, J_x, B_z and B_x are assumed small, substituting in eqn. 4.3 for θ_z in terms of θ_x derived from eqn. 4.2 and then substituting for T_x in eqn. 4.4 gives

$$\left(I + \frac{J_s^2\omega_s^2}{K}\right)\ddot{\theta}_x + B_s\dot{\theta}_x + WH\theta_x = F(t) \tag{4.5}$$

which, as was indicated earlier, shows that the effect of the gyro-stabiliser is to increase the effective inertia of the ship.

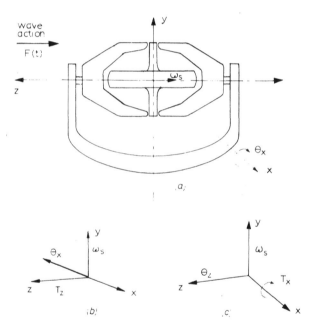

Fig. 4.20 *Arrangement of passive gyrostabiliser*

In the active system of gyrostabilisation Sperry used a sensor to detect the rolling of the ship. The sensor output operated, through a relay, a motor which caused the gyroscope to precess, thus generating a torque opposing the roll. The first sensor proposed by Sperry was a simple pendulum; to transit the motion of this sensor to the motor controller he used what he termed 'the phantom'. This is shown schematically in Fig. 4.21. Two contacts A and B were placed a small distance apart, one on each side of a contact C attached to the pendulum.

As the pendulum responded to the roll of the ship, it made contact with either A or B and completed the circuit to the motor D, so that the contacts were driven in the same direction as the pendulum; the

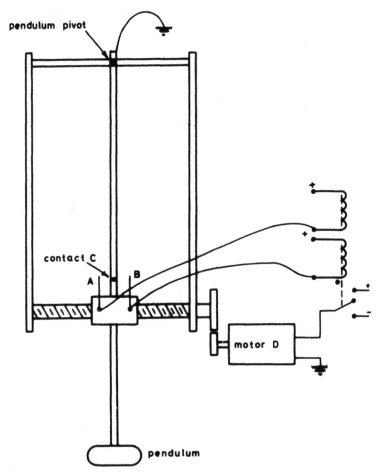

Fig. 4.21   *Principle of Sperry's phantom*
[Reprinted from Hughes, J. P.: *Elmer Sperry: inventor and engineer* (Johns Hopkins University Press, 1971) © Johns Hopkins University Press, 1971]

contact will thus follow the pendulum.[43] By the use of feedback Sperry obtained, in effect, proportional action from a relay system. The speed of the phantom was proportional to the rate of roll. A similar technique was to be used some years later by Bush on an early form of the differential analyser.[44]

In this scheme the follow-up motor also drove a drum controller for the motor; the drum controller was programmed to control the direction of precession and to increase the speed with the angle of roll. The restoring torque was therefore proportional to the angle of roll. Some time later Sperry seems to have realised that the restoring torque needs to be related to the rate of roll rather than angle of roll.[45] An undated sketch in Sperry's papers (Fig. 4.22) shows the system which was eventually adopted by the Sperry Gyroscope Company: the pendulum has been replaced by a small 'control gyroscope' which provides a simple way of determining rate of roll. The control gyroscope activates the contacts of a motor, thus causing the main gyroscope to precess about the z-axis at some fixed speed $\dot{\theta}_z$.

It was claimed in the 1930s that the control gyroscope fitted in the *Conte di Savoia* indicated the angular velocity of roll,[46] and that the rate of precession $\dot{\theta}_z$ was proportional to the rate of roll $\dot{\theta}_x$; hence $\dot{\theta}_z = K_p\dot{\theta}_x$. Substituting in eqn. 4.3 for $\dot{\theta}_z$ and then in eqn. 4.4 for T_x now gives an equation of motion for the stabilised ship of the form

$$I\ddot{\theta}_x + (B_s + K_p J_s \omega_s)\dot{\theta}_x + WH\theta_x = F(t) \tag{4.6}$$

which shows that the effect would be to increase the natural damping of the ship. There is, however, no evidence that the control gyroscope was ever adopted to indicate angular velocity. 'From all I know about the Sperry pilot gyro it only indicates the direction of the ship's angular velocity'[47] was Schilovsky's view. The Sperry system was in fact a 'bang-bang' control system which, simplified by ignoring T_m, J_2 and B_2, results in an equation of motion for the ship of the form

$$I\ddot{\theta}_x + B_s\dot{\theta}_x + \mathrm{sgn}\,(\dot{\theta}_x)K_m J_s \omega_s + WH\theta_x = F(t) \tag{4.7}$$

'The gyroscopic method, in spite of its initial success on small ships, became obviously impracticable when attempted on the larger scale of the steamship *Conte di Savoia*. The excessive initial cost of the installation, the considerable space required in the central part of the ship, the care and maintenance needed for proper upkeep of enormous gyroscopes, all created a definite objection to this method of stabilization' was Minorsky's comment in his paper describing experiments on a new method of stabilisation using active tanks.[48]

In 1890, A. Wilson patented a method of stabilisation based on the use of oscillating fins; at the beginning of the present century an unsuccessful attempt to use the technique was made by a Japanese inventor S. Motora. Between the wars the idea was taken up by Brown Brothers & Co. Ltd, and by William Denny & Bros. Ltd. The first successful Denny–Brown active fin stabiliser was installed in a

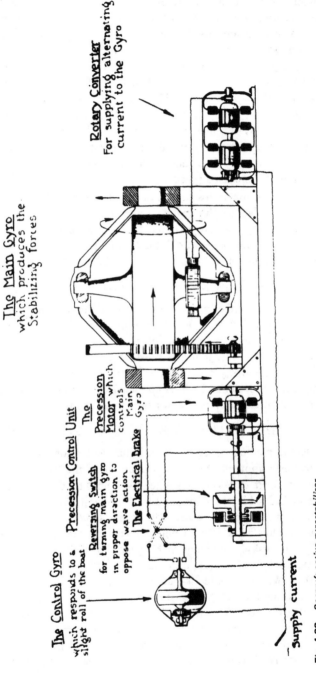

The Main Gyro
which produces the
stabilizing forces

Rotary Converter
For supplying alternating
current to the Gyro

The Control Gyro   Precession Control Unit

The
Precession
Motor which
controls
Main
Gyro

Reversing Switch
for turning main gyro
in proper direction to
oppose wave action

The Electrical Brake

which responds to a
slight roll of the boat

Supply current

Fig. 4.22   Sperry's active gyrostabiliser
[Reprinted from Hughes, J. P.: Elmer Sperry: inventor and engineer (Johns Hopkins University Press, 1971) © Johns Hopkins University Press, 1971]

crosschannel steamer, the *Isle of Sark*, in 1936.[49] In the first installations the movement of the fins was made proportional to the rate of roll, but as J. F. Allen explained

> A degree of anticipation may be introduced into this system [a gyroscopic system for determining angular velocity of roll] by mounting the gyro contacts on a carriage which slides against a mechanical resistance under the precessional force of the gyro, so that when the gyro begins to centre as the angular velocity of the ship dies out, the contacts are immediately reversed, and the fins consequently reverse, in anticipation of the next rolling impulse, and also partly offset the unsuitable time lag in the movement of the fins.[50]

This was, of course, an application of Sperry's 'phantom'. During the Second World War this method of stabilisation was further developed by J. Bell, at the Admiralty Research Laboratory, to include a continuous proportional, as well as a continuous derivative, signal.[51]

Automatic steering
In the 1870s Werner Siemens, concerned for the safety of the German nation, turned part of his attention and that of his firm, Siemens & Halske, away from the electric telegraph, dynamo-electric machines and arc lamps, to the development of weapons and, in particular, torpedo boats. One of his ideas was for the automatic steering of a torpedo boat, and trials on a boat loaned by the German Navy were carried out between 1872 and 1874. The rudder of the boat was turned by an electric motor which was operated by electromagnet relays; the control could be either by hand from a station on land connected to the boat by light cable or from the magnetic needle of a compass placed in the boat.[52]

With the development of steering engines during the latter quarter of the 19th century it is not surprising that attempts were made to connect the steering engine to the magnetic compass. One such attempt was made by A. B. Brown. Reports of the trials of his system indicate that it was unsuccessful because he 'was endeavouring to use deviation control [proportional control] alone, without check helm [derivative action], with the result that the ship followed a sinuous course, and the angle of oscillation increased with the speed.[53] Similar problems had occurred with the steering of torpedoes using the Obry gyroscope.[54]

It was in connection with work on improving the steering of torpedoes that Sir James B. Henderson introduced the idea of 'check helm' and was granted a patent in 1913 for an automatic steering device in which the control action was dependent both on the deviation from the course and the rate of change of deviation.[55] The basis of the invention was a constrained gyroscope used to measure the angular velocity of the ship, for which Henderson had been granted a secret patent in 1907. The constrained gyroscope was reinvented by Elphinstone and Wimperis in 1910 and described in a published patent.[56] However, the major contributions to the development of a practical automatic steering system were made by the Sperry Gyroscope Company.

Elmer Sperry began work on designing an automatic pilot, the gyropilot as it became known, in 1912; the basic patent was filed in 1914 and granted in 1920. The war interfered with development, and it was not until after the war that a practical gyropilot was produced. The decision to develop the gyropilot was made on commercial grounds; the Anschütz company was found to be working on an automatic ship's pilot, and the Sperry Company believed that if they could not supply an automatic ship's pilot as an addition to the gyrocompass they would lose ground in the gyrocompass market. Effort was put into developing the gyropilot, trials were carried out in April 1922, and in October of the same year the first permanent installation was made in the liner *Munargo*. By 1932 over 400 gyropilots were in service.[57] The gyropilot had been christened 'Metal-Mike' by the officers of the ship *Moffett*, which had been used for the trials, and the performance of 'Metal-Mike' seemed uncanny to many because it apparently had built into it the 'intuition' of an experienced helmsman. In a sense, of course, it had.

The basis of the gyropilot was an electric motor which, by means of a 3-position controller, could be rotated in either direction or held stationary. The motor was used to turn the helm, which then operated the rudder through the normal steering engines of the ship. Sperry had realised when he started work on the gyropilot in 1912 that control based on the sign of the deviation would not be satisfactory: a helmsman 'eases' the helm as the ship responds – his 'intuition' is to make the rudder angle *proportional* to the deviation. This Sperry achieved by using feedback from the helm, as shown in Fig. 4.23a. The input ϵ_e to the minor loop, the deviation from the set course, is represented in the gyropilot by the angle of rotation of a drum carrying positive and negative electrical contact strips separated by an insulating zone (Fig. 4.23b). A trolley slides over the drum, and, together, they form the 3-way switch used to operate the helm drive motor. The trolley is mounted on a platform which turns with the movement of the helm. The inner loop thus tries to keep the trolley aligned with the insulating strip on the drum and hence makes the position of the rudder proportional to the deviation.[58]

An experienced helmsman would also, Sperry observed, 'meet' the helm, that is, back off the helm and put it over the other way to prevent the angular momentum of the ship carrying it past the desired heading. To provide for this Sperry designed what he called an

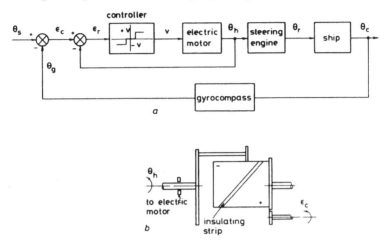

Fig. 4.23   *a*   Block diagram of Sperry's basic gyropilot
              b   Controller for basic gyropilot
              θ_s = set course
              θ_c = actual course
              θ_g = measured course
              θ_h = helm angle
              θ_r = rudder angle

'anticipator' and this device was included in the patent filed in 1914. The anticipation effect was achieved by introducing backlash into the helm position feedback linkage (Fig. 4.24). Thus when the sense of ϵ_c changes, that is, when the ship is no longer deviating from the course, but is beginning to swing back onto course, the inner loop is allowed to operate as a bang-bang controller and made a sudden change, of a limited amount, to the setting of the helm, before reverting to proportional action. Sperry argued that the amount of correction to be allowed should be proportional to the overshoot, 'the left over expression of the amplitude of yaw', which would occur in the absence of the anticipator'. He further argued that the overshoot would depend on the amplitude and period of the yawing motion.[59] The 'anticipator' therefore contained a mechanism, shown in simplified form in Fig. 4.25, which would automatically adjust the backlash according to the amplitude of the yawing motion.[60] The shears AB are separated by the action of the pin P, which follows the error signal ϵ_c; the shears close

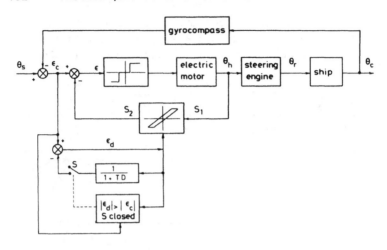

Fig. 4.24   *Block diagram of gyropilot with adaptive control*
I = time constant of spring–dashpot unit (Fig. 4.25). Other symbols
as in Fig. 4.23

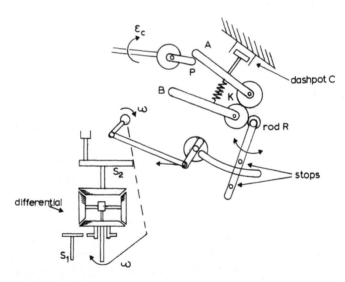

Fig. 4.25   *Schematic showing operation of Sperry's 'lost-motion' device*
[Simplified version of diagram in Hughes, J. P.: *Elmer Sperry: inventor
and engineer* (Johns Hopkins University Press, 1971) © Johns Hopkins
University Press, 1971]

up again slowly under the action of the spring K and dashpot C. The movement of the shears operates the rod R, which sets the angle through which the centre of the differential can turn freely; with the centre of the differential free no motion can be transmitted from shafts S_1 to S_2; with it locked the shafts are directly coupled. As can be seen from the overall block diagram (Fig. 4.24) the proposed system involved adaptive control.

It is not certain whether or not this full mechanism was included in the production gyropilots. In an article in *The Engineer* in 1937, following a discussion of the effect of the 'lost motion' (backlash), there appears the statement that 'In the Sperry gyropilot no provision is made for automatically varying the lost motion in the manner we have indicated [making it proportional to the amplitude of the deviation]. Its amount can be adjusted but only manually and solely for the purposes of meeting changes in the condition of the sea or ship.'[61] The lost-motion device was described by a Sperry engineer as 'a dodge. It is an unscientific dodge. It defies mathematical computation – at least it has, defied mine.'[62] It is perhaps for this reason that the even greater 'dodge' of automatically adjusting the lost motion was not used. Had Elmer Sperry still been fully involved in the development of the gyropilot, dodge or no dodge, amenable to analysis or not, there is little doubt that the device would have been used and made to work.

Aeroplane stability and control

> Apart from the fact that movable parts are liable to get out of order, it must be remembered that they increase the number of degrees of freedom of the machine, this further adding to the number of conditions which have to be satisfied for stability – a number quite large enough already. I anticipate that the successful aeroplane of the future will possess inherent not 'automatic' stability, movable parts being used only for purposes of steering.'
>
> G. H. Bryan, 1911

The pioneers of aviation, Lilienthal, Maxim, Lanchester, Pilcher, Chancute and Langley, attempted to build aircraft which had inherent stability.[63] Lilienthal's success in achieving inherent stability ultimately killed him: the stability which he achieved was with respect to the air mass not the ground, and on his last flight he had insufficient control power to counteract a gust disturbance.[64] The Wright brothers rejected the dogma of inherent stability: 'We therefore resolved to try a fundamentally different principle. We would arrange the machine so that it would not tend to right itself.'[65] The pilot was required to 'fly' their aircraft; to enable him to do so he was provided with powerful control surfaces. However, to successfully operate these controls, the pilot needed to be able to sense the motion of the aircraft; in good conditions and with visual contact with the ground this he could do, but in fog or at night, he lacked the necessary visual information.

The early practitioners saw two possibilities: provide the pilot with instruments to indicate the behaviour of the aircraft or provide automatic control. There were, as Elmer Sperry explained, some compelling reasons for adopting the latter approach:

> With the present machines very long flights are nearly beyond the endurance of the aviator [because of fatigue in maintaining stability] . . . . The automatic control of stability of the heavier-than-air machines will do much to decrease the growing list of fatalities . . . . The automatic control of stability will be especially valuable to the military use of the aeroplane, as it will make it possible to fly in almost any condition of weather . . .[66]

In 1897 Sir Hiram Maxim had made provision for automatic pitch control on his experimental steam-powered aeroplane; a pendulous gyroscope, used to detect pitching motion, was coupled by a steam-operated servovalve to the elevators.[67] In 1909 Elmer Sperry proposed using a passive gyroscope, similar to Schlick's ship stabiliser, in the Beach monoplane, and he also tried to sell the idea to the Wright brothers. Following tests in the spring of 1910, Sperry lost interest in the passive stabiliser, because he had begun to realise that a stabiliser needed to act before the aircraft or ship gathered momentum, that it needed to be active.[68]

There were numerous attempts during the next few years to develop active automatic stabilisers for both longitudinal and lateral motion. Table 4.1 shows some of the variables which were measured and the instruments used for measurements (the notion is defined in Fig. 4.26), and Tables 4.2 and 4.3 list the various feedback variables and actuators used in aircraft stabilisers.[69]

The outstanding early contributions came between 1912 and 1914 from Elmer Sperry and his son Lawrence.[70] In Paris in 1914 Lawrence Sperry demonstrated the Sperry autostabiliser to the jury of the Union for Safety in Aeroplanes and numerous military attachés. In a superb publicity stunt Lawrence stood with his hands off the controls while his mechanic walked along the wing of the aircraft: the aircraft maintained straight and level flight.

For this system the Sperrys used four gyroscopes mounted to form a stabilised reference platform; a train of electrical, mechanical and

Table 4.1 *Feedback signals and instruments for aircraft stabilisers*

	Parameter	Symbol	Instruments used
Longitudinal motion	1. Speed	v	Anemometric plate, venturi
	2. Incidence	$\alpha = -\omega/u$	Vane placed in wind layer
	3. Inclination to horizontal	θ	Free gyroscope suspended from its centre of gravity
	4. Angular pitching velocity	q	Gyroscope producing precession couple
	5. Gravity plus acceleration, x-axis	$g \sin \theta + \dfrac{du}{dt}$	Pendulum, accelerometer along x-axis of aircraft
	6. Gravity plus acceleration, z-axis	$g \cos \theta + \dfrac{dw}{dt}$	Accelerometer along z-axis
	7. Lift	αu^2	Inclined plate, manometer indicating difference between pressure at two suitable points above and below wing
	8. Rate of climb	$u \sin \theta$	Variometer (climb indicator)

Table 4.1   *(Continued)*

	Parameter	Symbol	Instruments used
Lateral motion	1. Angle of slideslip	$j = v/u$	Vane with vertical axis
	2. Bank angle	ϕ	Free gyroscope suspended from its centre of gravity
	3. Azimuth	ψ	Free gyroscope, magnetic compass, earth-indicator compass
	4. Angular velocity of rolling	p	Gyroscope producing a precession couple
	5. Angular velocity of yaw	r	Gyroscope, difference in linear speed of wing tips
	6. Gravity plus acceleration along y-axis	$g \sin \phi + \dfrac{dv}{dt} + V_r$	Pendulum in ZOY plane, accelerometer along y-axis

pneumatic components detected the position of the aircraft relative to the platform and applied correction signals to the aircraft control surfaces. The stabiliser operated for both longitudinal motion (pitch) and lateral motion (roll); in both cases a minor feedback loop from the position of the control surface was incorporated. Since the elevator angle controls the rate of change of pitch angle, the feedback of elevator angle is equivalent to velocity feedback, and hence provides damping. The system was normally adjusted to give an approximately deadbeat response to a step disturbance. The incorporation of derivative action — the equivalent of 'meeting' the helm — was based on Sperry's intuitive understanding of the behaviour of the system, not on any theoretical foundations. The system was also adaptive. The output of an anemometer, used to measure the air speed, shifted the fulcrum of the levers which operated the control surfaces, thus adjusting the gain to match the speed of the aircraft.

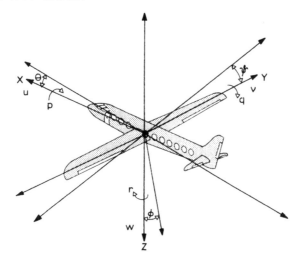

u,v,w components of velocity in OX, OY, OZ directions

p,q,r angular velocities about OX, OY, OZ directions

θ, ψ, ϕ angle of inclination of OX to horizontal ; angle

  of azimuth ; angle of banking.

Fig. 4.26   *Nomenclature for aircraft motion*

Before autostabilisers became commercially available, the First World War began. War changed the requirements: flights were now of short duration, fighters had to be highly manoeuverable and pilots

Table 4.2 *Longitudinal stabilisers*

Control parameters	Symbol	Inventor and date	Actuating means
Air speed	v or u	Budig, 1912 Eteve, 1914	Direct mechanical connection
Incidence angle	α	Eteve, 1910 Constantin S. T. Aé	Direct mechanical Direct mechanical Electric
Inclination to the horizon	θ	Regnard, 1910 Sperry, 1912 RAE, 1927 Smith Sperry, 1932 Pollock-Brown	Electric type of servo Air-turbine-driven clutch servo Pneumatic servo Pneumatic servo Hydraulic servo (oil) Hydraulic servo (oil)
Angular pitching velocity	q	Girardville, 1910	Direct mechanical connection
Direction of apparent gravity	$g \sin \theta + \dfrac{du}{dt}$	Moreau, 1912	Electric-motor-driven clutch servo or direct
Direction of apparent gravity + air speed	$g \sin \theta + \dfrac{du}{dt} + u$	Dantre, 1911 Askania Mazade SECAT	Compressed air Compressed air Compressed air Mechanical with electro-magnetic clutches

Table 4.2 *(Continued)*

Control parameters	Symbol	Inventor and date	Actuating means
Apparent gravity + acceleration along OZ axis	$g \cos \theta + \dfrac{dw}{dt}$	Doutre, 1913 Boykow	Compressed air
As above + lift	$g \cos \theta + \dfrac{dw}{dt}$ $+ \alpha u^2$	Gianoli	Aerodynamic servomotor
Airspeed and pitching velocity	u and q	Boykow, 1928 Siemens	— Hydraulic servo (oil)
Airspeed and inclination	u and θ	Marmonier, 1909	—

Table 4.3 *Lateral stabilisers*

| Control parameter | | Inventor/manufacturer | Method of control |
Rudder	Ailerons		
	$j = v/u$	Constantin	Direct
	$g \sin \phi + (dv/dt) + V_r$	Gianoli, 1933	Aerodynamic servomotor
$r + g \sin \phi + (dv/dt) + V_r$	$g \sin \phi + (dv/dt) + V_r$	Gianoli, 1935	Aerodynamic servomotor
ψ		Mazade-Aneline 1922	Compressed air
ψ		Askania, 1927	Compressed air
ψ	$g \sin \phi + (dv/dt) + V_r$ and j	SECAT	Electric motor with electrically controlled clutches
ψ	$g \sin \phi + (dv/dt) + V_r$ and r	SECAT	Compressed air
ψ	ϕ	Sperry, 1912	Hydraulic servo (oil)
		Pollock-Brown	Hydraulic servo (oil)
		Smith	Compressed air
ψ and r	$g \sin \phi + (dv/dt) + V_r$ and r	Siemens, 1927	Hydraulic servo (oil)
		Siemens, 1932	
ψ		RAE, 1927	Pneumatic servo

were intensely proud of their flying skill; the pilot was firmly within the control loop. The emphasis was on the provision of a range of instruments to help the pilot, such as turn indicators, artificial horizons and slip indicators, and not on autostabilisation. There was one area, however, where stabilisers were still required, and that was for bombing.

In England, Bertram Hopkinson,* Tizard† and Lindemann‡ all contributed to the development of a bombing system. It was based on the use of a Sperry stabiliser with the addition of a directional gyroscope and a telescopic bombsight. As part of the bombing system, the pilot, after determining the drift, steered by the directional gyroscope; it was therefore proposed that the directional gyroscope, once set, should be linked by servomotors to the rudder and ailerons to steer the course automatically. Lawrence Sperry, who in 1915 visited England to assist the Sperry engineers based at Upavon, took the idea of automatic steering back to the USA, recommending to his father that the company concentrate on the development of 'an azimuth stabiliser'. This coupled with the normal Sperry stabiliser could provide the autopilot necessary for pilotless aircraft and for the aerial torpedo in which Elmer Sperry had become interested, a device which necessitated extensive development of automatic control systems.[71]

Aircraft autopilots

After 1918 considerable interest in radio-controlled aircraft developed, with a consequent interest in autopilots. This work was largely carried out by the military authorities, in Britain at the Royal Aircraft Establishment (RAE), in the USA by the Naval Research Laboratory. Information about this work was consequently restricted. Meredith and

* Bertram Hopkinson, Professor of Mechanism and Applied Mechanics at the University of Cambridge, joined the Royal Engineers in 1914; he was put in charge of the design and supply of bombs, bomb gears, guns and ammunition for the Department of Military Aeronautics in 1915. He was killed in an aircraft accident in August 1918.

† Henry Tizard (later Sir Henry) (1885–1959) scientist and Fellow of Oriel College, Oxford, volunteered at the beginning of the war and was commissioned into the Royal Garrison Artillery. During the early part of 1915 was transferred to the Royal Flying Corps to provide scientific assistance at the Central Flying School. After the war, he became, in various capacities, the leading government advisor on scientific matters in England [see the biography by Clark, R. W.: *Tizard* (London, 1965)].

‡ Frederick Alexander Lindemann (later Viscount Cherwell) (1886–1957) was born in Baden-Baden, but lived most of his life in England. During the early part of First World War he was a civilian scientist at Farnborough, working on bombsights. During the Second World War he was Chief Scientific Advisor to Churchill [see the biography by Lord Birkenhead: *The Prof. in two worlds: the official life of Professor F. A. Lindemann, Viscount Cherwell* (Collins, London, 1961)].

Cooke of the RAE were allowed to patent various autopilot developments in 1926 and 1927, but only after all references which might
indicate that the apparatus could be used for a pilotless aircraft had
been deleted.[72] In Germany prior to 1930 a few units were produced
by the Askania Company and by the Siemens Company, both working
in conjunction with Johann Maria Boykow. From 1928 onwards the
Siemens Company, because of the unsatisfactory nature of the Boykow
system, proceeded with its own development, directed by Eduard
Fischel. Fischel developed an electro-hydraulic system which provided
control of heading as well as stabilisation in pitch and roll.[73]

The 1930s saw the development of commercial autopilots: Wiley Post
insisted on using a Sperry autopilot prototype for his round-the-world
flight in 1933.[74] In England the Smith Instrument Company were
offering a commercial version of the RAE autopilot, and a commercial
version of the Siemens autopilot could also be purchased. Military
autopilots were also being developed by the RAE, and the Siemens,
Askania, Sperry Gyroscope, General Electric and Honeywell Companies.
The basis of all the autopilots developed was some form of stabilised
platform and a gyrocompass, but there were considerable differences
in the methods used to 'pick off' the signals and operate the various
actuators. The early British systems were pneumatic and considerable
difficulties occurred through small dust particles jamming valves,
whereas the American and German designers tended to use hydraulic
or mechanical systems. Gradually electrical components were introduced, but wholly electrical units were not produced until after the
Second World War. The importance of the development of autostabilisers and autopilots to control engineering is in the development
of components and in the idea of a control circuit combining several
signals. The systems were not designed on the basis of any control
theory and their general complexity, involving a mixture of electrical,
mechanical and pneumatic (or hydraulic) components, as well as non-
linearities, was not conducive to the development of any theory.

Theory of automatically steered bodies

> If, therefore, accurate steering is nothing more than a special kind
> of timing of the rudder complicated by the inertia of the body to
> be steered, we may expect to be able to establish analytically what
> kind of timing must be adopted in order to reach the best possible
> conditions for directional stability of the body to be steered on
> its course.

N. Minorsky, 1922

Nicholas Minorsky, born 24th September 1885 in Karcheva, Russia, was educated at the Naval School in St. Petersburg [Leningrad] and the University of Nancy. He taught at the Imperial Technical School in St. Petersburg from 1911 until 1914, when he joined the Russian Navy. In June 1918 he emigrated to the USA.[75] During his period of service with the Russian Navy, Minorsky studied the problem of the automatic steering of ships. He realised that any automatic steering system must take into account not just the angular deviation, but also the rate of change of the deviation. By 1918 he had devised a control scheme and an instrument, a gyrometer, to measure the angular velocity of yaw of a ship. He had even, in 1916, carried out experiments to determine the sensitivity of the eye of experienced helmsmen in detecting angular velocity as a check on the minimum requirement for the sensitivity of his gyrometer.[77] For his first four years in the USA Minorsky worked as an assistant to C. P. Steinmetz at the General Electric Company, Schenectady, before persuading, in 1922, the US Navy to test his system of automatic steering. The system was installed in the battleship USS *New Mexico* and trials were carried out during 1923 'with promising results'. The system was never developed and in 1930 Minorsky disposed of his patents on the subject to the Bendix Aviation Company.[78]

Minorsky's major contribution to the subject of automatic steering was not, however, the development of a practical device but a clear and concise theoretical analysis of automatic steering and the recommendation of the use of what would now be called '3-term control'. This analysis was presented in a paper 'Directional stability of automatically steered bodies' published in 1922.[79] Like Sperry and Henderson, Minorsky was well aware of the 'anticipation' used by a good helmsman, but, as his experiments on the ability of the helmsman to judge angular velocity show, he was aware that this 'anticipation', this 'intuition' as it was frequently called, was based on the helmsman making and using measurements other than simply deviation from the set course. What he also realised was that a good helmsman was not only able to make these additional measurements, but was able to make use of them at the right time. In addition Minorsky recognised that it was only necessary to consider the stability of motion of an automatically steered ship for *small deviations* from the desired course, a procedure 'which not only simplifies considerably the analytical solution of the problem but gives the only solution of practical interest, since only small deviations are possible, if the steering is to be accurate; if there is no stability for small deviations, it means that there is no stability in general since the ship will be continuously deviated from her course'.[80] By considering small deviations Minorsky was able to linearise the system about an operating point and he obtained an equation of motion for the steered ship,

$$\frac{A d^2 \alpha}{dt^2} + \frac{B d\alpha}{dt} + K\rho = D \qquad [2]^* \qquad (4.8)$$

where A is the moment of inertia of the ship, B is the viscous damping coefficient, K is a constant, α is the angular deviation from the set course, ρ is the angle of the rudder and D is the disturbance force.

Having obtained eqn. [2] he continues, 'The problem is completely determined if the relation of the angle of rudder ρ as a function of α and its time-derivatives is given .... Several methods of regulation are possible; we may mention two fundamental classes according to whether the regulation affects the angle of the rudder or the rate of change of this angle.'[81]

These methods of regulation were expressed by Minorsky in the form of simple linear relationships:

$$\rho = m\alpha + n\frac{d\alpha}{dt} + \frac{d^2\alpha}{dt^2} \qquad [X] \text{ first class} \qquad (4.9)$$

$$\frac{d\rho}{dt} = m_1\alpha + n_1\frac{d\alpha}{dt} + p_1\frac{d^2\alpha}{dt^2} \qquad [XX] \text{ second class} \quad (4.10)$$

$$\frac{d^2\rho}{dt^2} = m_2\alpha + n_2\frac{d\alpha}{dt} + p_2\frac{d^2\alpha}{dt^2} \qquad [XXX] \text{ third class, and so on} \qquad (4.11)$$

He then solved the differential equations arising from considering the application of his 'first class' of control with each term $m\alpha$, $n d\alpha/dt$; $p d^2\alpha/dt^2$ used independently and then in combination, concluding that this 'first class' of controller 'is efficient from the standpoint of damping and the effect of a disappeared disturbance, but will not eliminate the effect of a steadily acting disturbing torque, such, for instance, as a steady wind.[82]

Minorsky, therefore, turned to his 'second class' of controller, which gave rise to a system described by

$$A\frac{d^2\alpha}{dt^2} + B\frac{d\alpha}{dt} + K\int (m\alpha + n\alpha' + p\alpha'')dt = D \qquad (4.12)$$

By differentiating with respect to time and rearranging he found that

---

* The letters and numbers in square brackets are the notation used by Minorsky in his paper.

this equation reduces to

$$A \frac{d^3\alpha}{dt^3} + (B + Kp) \frac{d^2\alpha}{dt^2} + Kn \frac{d\alpha}{dt} + Km\alpha = \frac{dD}{dt} \qquad (4.13)$$

and, for a steadily acting disturbance,

$$\frac{dD}{dt} = 0$$

Hence Minorsky obtained 'the remarkable result that such a disturbance has no influence upon the performance of the device.'[83] He also noted that 'A complete solution of the auxiliary equation of the third degree is not necessary ... M. A. Blondel[84] has shown that by applying the so-called Hurwitz theorem of Analysis, the "stability of roots" of an algebraic equation of the nth degree can easily be established'.[85] The stability conditions are thus found to be

(1) $B + Kp > 0$

(2) $(B + Kp)Kn - AKm > 0$

(3) $Km > 0$

The control scheme recommended by Minorsky, his second class of control,

$$\frac{d\rho}{dt} = m_1\alpha + n_1 \frac{d\alpha}{dt} + p_1 \frac{d^2\alpha}{dt} \qquad (4.14)$$

provides 3-term control since integrating eqn. 4.14 gives

$$\rho = \int m_1\alpha \, dt + n_1\alpha + p_1 \frac{d\alpha}{dt} \qquad (4.15)$$

There was, however, some confusion about the system. For example, Minorsky himself expressed some surprise [his italics] 'that the gyrometer action does not *add itself* to the resistance of the medium B'.[86] It was not realised that because of the integration implicit in controlling rudder velocity the gyrometer was in effect being used to measure angular deviation.

The first trials of Minorsky's systems in the USS *New Mexico* were carried out with the controller operating from deviation and velocity signals (deviation and check helm), but Minorsky had arranged the equipment to control the *angular velocity* of the rudder, not the angle. The control was thus proportional and integral, and during the trials the system gave a sustained yaw of $\pm 6°$; increasing the amount of check helm, that is proportional action, reduced the yaw to ± 2°, but Minorsky was unable to get any further improvement. He therefore added acceleration control, that is, derivative action, and the yaw was now reduced to ± 1/6°, better than most helmsmen can achieve. In a final section of his paper Minorsky considered the effect on the stability of the steering system of time delays in the measurement of α, α' and α''.

This contribution of Minorsky to the theory of automatic control, the first theoretical analysis of the various types of controller, received little immediate attention. Gates, in a classified report on aircraft steering in 1924 alluded to the use of derivative action in the steering of large ships, but apparently was not aware of Minorsky's work.[87] Like Minorsky, he pointed out that stability alone was not enough:

> In the problem of an automatic control it is not merely necessary that the transition motion involved in a change of direction should be characterised by reasonably short periods and high rates of decay. A satisfactory control must satisfy the further condition that the amplitudes of the auxiliary motions . . . consequent on a unit exercise of the control should not transcend limits which are more or less fixed by practical considerations . . . Evaluation of the amplitudes of the motion as well as its exponential factors is therefore necessary.[88]

He also attempted to analyse the interactive effects which arise in aircraft control: 'The imperfections of the simple gyro-rudder control . . . arise, as might be expected, mainly from a lack of rolling control . . . The subsequent calculations were directed towards reducing the amplitudes and increasing the damping of the long oscillation [that of banking and sideslip] without altering materially the already satisfactory motion in turn.'[89] A number of methods for improving control were considered, but attempts were handicapped by the need for extensive numerical calculation for each case considered. He concluded that there was no simple way of reducing the interaction: 'The failure of such systems as have been considered must be due to the fact that none of the simple and obvious ways of working the control imitates the complex actions of a pilot at all closely.'[90]

Computational difficulties seem to have prevented development of an interest in the analysis of aircraft control systems. Melvill-Jones in 1935 commenting on the neglect of the simpler theory of the dynamics of the uncontrolled aircraft says:

> In spite . . . of the completeness of the experimental and theoretical structure . . . it is undoubtedly true, at the time of writing [1935], calculations of this kind are very little used by any but a few research workers. It is in fact rare for anyone actually engaged

upon the design and construction of aeroplanes to make direct use of [such] computations . . . , or even to be familiar with the methods by which they are made . . . In my opinion it is the difficulty of computation . . . which has prevented designers of aeroplanes from making use of the methods.[91]

The computation difficulty should not be underestimated; the dynamics of the basic aircraft without control are represented by a 4th-order differential equation and the addition of control increases the order to five or higher.

Conclusions

Necessity is the mother of invention, so the reverse would also seem true, each development in science and engineering necessitating the development of some device or devices to increase the efficiency of the prime development.

E. Sperry: 'Engineering applications of the gyroscope', *J. Franklin Inst.*, 1913, **165**, pp. 447–482

So wrote Elmer Sperry in 1913. His words provide an apt summary of the development of servomechanisms to that date and in fact beyond, up to the beginnings of the Second World War. The large ship necessitated the steering engine, which, in its turn, raised many mechanical problems and what we would now recognise as control problems. In a similar way the aeroplane brought into being a whole range of control and instrumentation problems, each in their turn raising other problems: the design of pneumatic valves, of hydraulic valves and the design of electrical transmitters. This was the period of the inventor, advances were made by cut and try, by intuition, by obtaining a 'feel' for the problem. The stabilisers, the gyropilots, the autopilots all exhibited control problems in their early stages of development. They were sluggish or they caused hunting. Solutions to these problems were obtained by considering, for example, what the human helmsman did, he 'eased' the helm, and he 'met' the helm, and as far as possible incorporating this action in the automatic control.

As with the engine governor, gradually an awareness developed the servomechanisms — controllers — did not work instantaneously, that time lags were of the essence. Hele-Shaw measured the lags present in a steam steering engine and in a hydraulic steering engine; the Sperry Company engineers discovered that time lags in conventional steering systems seriously affected their gyropilot, and they reduced the lag by using electrical transmission. The causes of lag were clearly stated by Minorsky: 'the delay is due to the transmission system possessing a certain amount of inertia, either mechanical or electromagnetic, lost motion, viscosity and similar causes. By perfecting the transmitting means the lag may be reduced, but never totally eliminated. Any transmission system can thus be characterised in terms of its lag measured, for example, in seconds.'[92] Kelvin had recognised that there is a time lag in cable transmission, Stodala had characterised governor-turbine behaviour in terms of time constants; in 1922 Minorsky introduced similar ideas into general servomechanisms.

Minorsky's work had little immediate impact; it was an achievement to show theoretically that good automatic steering required 3-term control, but there was a considerable problem in designing and building reliable apparatus to measure and combine the three terms. In this area Minorsky was in competition with a skilled inventor (Sperry) backed by sound engineers, and also with Anschütz Company, which had a 20-year start and time to build up a good engineering team.

Application of theory was also inhibited by the problems of computation. This was particularly evident in the area of aircraft control, although if the problems had been considered particularly pressing the computations would have been carried out.

References and notes

1 URE, A.: *The philosophy of manufacturers: or an exposition of the scientific, moral and commercial economy of the factory system of Great Britain* (London, 2nd edn. 1835), p. 18

2 SADI CARNOT: *Réflections sur la puissance motrice du feu er sur les machines* (Paris, 1824), quoted from KLEMM, F.: *A history of western technology* (George Allen & Unwin, London, 1959), pp. 275–276

3 Steam-powered ships had been used before this date; Robert Fulton's ship *Clermont* was first used to convey passengers between New York and Albany in 1807 and the British Navy accepted its first steamship in 1820. They were, however, small ships; Brunel changed the scale of shipbuilding and his ships mark the beginning of an era

4 Quoted from ROLT, L. T. C.: *Isambard Kingdom Brunel* (Penguin Books, Harmondsworth, 1970), p. 249

5 *ibid*., p. 324

6 *ibid*., p. 321. It is perhaps an indication of the vision of Brunel that members present at a meeting of the Institution of Mechanical Engineers in 1854 could not foresee any practical use of a gyroscope. See PARSONS, R. H.: *History of the Institution of Mechanical Engineers* (I Mech. E, London, 1947), pp. 110–111

7 *Engineering*, 1874 18, pp. 267–268, 286, 288–289, 475; 1875, 19, pp. 226–228, 291–293, 329–332, 407–408. Unfortunately, Bessemer had an incorrect understanding of the principle of the gyroscope. In the arrangement which he

designed the suspension was such that the only effect was that of a pendulum constrained to swing athwartships; see GRAY, J. M. *in Engineering* 1874, **18**, p. 307. The scheme which had been carried out at Bessemer's own expense was quickly abandoned

8 TOWER, B.: 'An apparatus for providing a steady platform for guns etc. at sea', *Trans. Inst. Naval Arch.*, 1889, **30**, pp. 348–361

9 THORNYCROFT, J. I.: 'Steadying vessels at sea', *Trans. Inst. Naval Arch.*, 1892, **33**, pp. 147–159; see also HUGHES, T. P.: *Elmer Sperry: inventor and engineer* (Johns Hopkins, Baltimore, 1971), pp. 114–115

10 GRAY, J. MacFarlane: 'Description of the steam steering engine in the *Great Eastern* steamship', *Proc. I Mech. E*, 1867, p. 267

11 MAYR, O.: *Feedback mechanisms* (Smithsonian Institute Press, Washington, 1971), pp. 93–94

12 CONWAY, H. S.: 'Some notes on the origins of mechanical servomechanisms', *Transactions of the Newcomen Society*, 1953–1955, **29**, p. 57

13 British Patent 3321, 1866, J. MacFarlane Gray

14 GRAY, J. MacFarlane: contribution to the discussion of the paper by Brown, A B.: 'On the application of a system of combined steam and hydraulic machinery to the loading, discharging and steering of steamships', *Trans. Inst. Naval Arch.*, 1890, **31**, p. 155. Quoted from CONWAY: *op. cit.*, p. 59

15 British Patent 2476, 1868, Joseph Farcot; FARCOT, J. J. L.: *Le servo-moteur ou moteur asservi* (Baudry, Paris, 1873)

16 *ibid.*, p. 4.: 'To put any motor or engine under the absolute control of an operator by the movement of his hand directly or indirectly on the control member of the motor, so that the two go, stop, go back and forward together, the motor following step by step the finger of an operator, imitating as a slave every moment. We believe it necessary to give a new name and characteristic for this new engine and we have called it servo-motor or slave motor' (translation is from Conway: *op. cit.*, p. 50)

17 *ibid.*, p. 2: 'being entirely unstable in its movement, and passing instantly, without cause, or for the slightest opening of the slide valve from one extreme of the position to the other.'

18 *ibid.*, p. 48

19 CONWAY: *op. cit.*, p. 61

20 *The Times*, 15 October 1878

21 INGLEFIELD, E. A.: 'On the hydraulic steering gear, as being fitted to HMS *Achilles*', *Trans. Inst. Naval Arch.*, 1869, **10**, pp. 92–100

22 *ibid.*, p. 94

23 BROWN, A. B.: British Patent 1018/1870; quoted from Conway *op. cit.*, p. 61

24 BROWN, A. B.: British Patent 2253/1871

25 MACNEIL, I.: 'Hydraulic power transmission', *Chartered Mechanical Engineer*, July, 1963, p. 359; 'Hydraulic power transmission: the first 358 years', *Transactions of the Newcomen Society*, 1974–1976, **47**, p. 154

26 LINCKE, F.: 'Das mechanische relais', *Zeitschrift des Vereins deutscher Ingenieure*, 1879, pp. 510–524, 578–616

27 HOWARD, H. S.: 'Hydraulic steering gears', *J. Am. Soc. Naval Eng.*, 1922, **34**, pp. 259–261

28 MARTINEAU, F. L.: reply to discussion on paper by HELE-SHAW, H. S., and MARTINEAU, F. L.: 'Steering-gear experiments on the turbine yacht *Albion*', *Trans. Inst. Naval Arch.*, 1915, **53**, p. 217

29 *ibid.*, p. 211

30 MAYR: *Feedback mechanisms*, p. 67

31 ARMSTRONG, G. E.: *Torpedoes and torpedo vessels* (Bell, London, 1901, 2nd edn.), p. 41

32 HENDERSON, J. B.: 'The automatic control of the steering of ships and suggestions for its improvements', *Trans. Inst. Naval Arch.*, 1934, 76, p. 21

33 This account is largely drawn from a series of articles by Commander Peter Bethell published in *Engineering* in 1945. BETHELL, P.: 'The development of the torpedo', *Engineering* 1945, **159**, pp. 403–405, 442–443; 1945, **160**, pp. 4–5, 41–43, 301–303, 341–344, 365–367, 529–531; 1946, **161**, pp. 73–74, 121–122, 169–170, 242–244. See also MAYR: *Feedback mechanisms*, pp. 96–101

34 COLLINS, F.: *Scientific American*, 1907, **16**: 'The gyroscope as a compass', p. 294; 'New uses for an old device – the gyroscope railroad train', p. 406; 'The Brennan gyroscope monorail', pp. 449–450; 'Practical tests of the Schlick gyrostat for ships', p. 494

35 HUGHES: *op. cit.*, pp. 106–107

36 MAYR: *Feedback mechanisms*, p. 101; Hughes: *op. cit.*, pp. 130–131

37 HUGHES: *op. cit.*, p. 108; patent was filed on 2 December 1907

38 *ibid.*, p. 107

39 *ibid.*, p. 111

40 *ibid.*, p. 134

41 RICHARDSON, K. I. T.: *The gyroscope applied* (Hutchinsons, London, 1954), p. 54

42 US Patent, filed 21 May 1908 (434 048), 'Ship's gyroscope', issued 17 August 1915 (Patent 1 150 311); quoted from HUGHES: *op. cit.*, p. 112

43 HUGHES: *op. cit.*, pp. 113–114

44 BUSH, V., GAGE, F. D., and STEWART, H. R.: 'A continuous integraph', *J. Franklin Inst.*, 1927, **211**, pp. 63–84

45 RAWLINGS, A. L.: discussion on paper by SCHILOVSKY, P.: 'Preliminary calculations of the sizes of gyroscopes required to stabilise a ship', *Trans. Inst. Naval Arch.*, 1933, 75, p. 165

46 *ibid.*, p. 165; see also HAIGH, B. P.: discussion on paper by SCHILOVSKY P.: *op. cit.*, p. 166

47 *ibid.*, p. 170

48 MINORSKY, N.: 'Experiments with activated tanks', *Trans. ASME*, 1947, 69, pp. 735–747

49 RICHARDSON: *op. cit.*, pp. 148–149

50 ALLAN, J. F.: 'The stabilization of ships by activated fins', *Trans. Inst. Naval Arch.*, 1945, 87, p. 141

51 RICHARDSON: *op. cit.*, p. 148

52 SIEMENS, W.: *The scientific and technical papers of Dr. Werner Siemens* (Murray, London, 1895), pp. 334–337

53 HAIGH, B.: contribution to discussion on paper by HENDERSON, J. B.: *op. cit.*, pp. 30–31

54 BETHEL: *op. cit.*, 1945, **160**, p. 303

55 HENDERSON, J. B.: British Patent 29661/13; see HENDERSON: *op. cit.*, p. 22

56 *ibid.*, p. 24

57 HUGHES: *op. cit.*, pp. 278–279

58 In later systems the 'trolley' was replaced by an E-core inductive pick-up; see RICHARDSON: *op. cit.*, p. 71

59 HUGHES: *op. cit.*, p. 282

60 A copy of the drawing given in the patent specification can be found in HUGHES: *op. cit.*, p. 281

61 'The principles and practice of automatic control, no. 6', *The Engineer*, London, 1937, **163**, p. 237

62 RAWLINGS: 'Contribution to discussion of paper by HENDERSON: *op. cit.*, p. 28

63 DRAPER, C. S.: 'Flight control', *J. Roy. Aero. Soc.*, 1955, **59**, pp. 451–477

64 McRUER, D., and GRAHAM, D.: 'A historical perspective for advances in flight control systems', *AGARD Advances in Control Systems*, No. 137, 1974, p. 2–1

65 WILBUR WRIGHT: lecture given to Western Society of Engineers, 13 September 1901; quoted from DRAPER, C. S.: *op. cit.*, p. 463

66 SPERRY, E.: 'Engineering applications of the gyroscope', *J. Franklin Inst.*, 1913, **175**, pp. 447–482; quoted from HUGHES: *op. cit.*, p. 174

67 HUGHES: *op. cit.*, p. 187

68 *ibid.*, pp. 176–177

69 The tables have been assembled from information given by HAUS, F.: 'Automatic stabilization', *NACA Technical Memorandum*, 802, 1936, (a translation of 'La stabilisation automatique', *L'Aeronautique*, Oct. 1935, pp. 81–87; Jan. 1936 pp. 1–6; Feb. 1936 pp. 17–23), and by McRUER, D.: GRAHAM, D.: *op. cit.*, pp. 21–27

70 For a detailed account of this work, see HUGHES: *op. cit.*, pp. 181–200

71 HUGHES: *op. cit.*, pp. 261–262

72 PRO AVIA 8 209/Pt I. Automatic controls – disclosure of technical particulars

73 A brief account of German autopilot developments has been given by OPPELT, W.: 'A historical review of autopilot development, research, and theory in Germany', *J. Dynamic Systems, Measurement, and Control, Trans. ASME*, series G, 1976, **98**, pp. 215–223

74 DRAPER: *op. cit.*, p. 471

75 FLUGGE-LOTZ, I.: 'Memorial to N. Minorsky', *IEEE Trans.*, 1971, **AC-16**, pp. 289–291

76 *The Engineer*, 1937, **163**, p. 322

77 MINORSKY, N.: 'Directional stability of automatically steered bodies', *J. Am. Soc. Naval Eng.*, 1922, **34**, p. 284

78 *The Engineer*, 1937, **163**, p. 322–323, 352–353, and MINORSKY, N.: 'Automatic steering test', *J. Am. Soc. Naval Eng.*, 1930, **42**, pp. 285–310 give details of the steering mechanism

79 MINORSKY: *op. cit.*, 1922, pp. 280–309

80 *ibid.*, p. 283

81 *ibid.*, p. 289

82 *ibid.*, p. 299

83 *ibid.*, p. 300

84 BLONDEL, M. A., *Journal de Physique*, 1919

85 MINORSKY: *op. cit.*, 1922, pp. 301–302

86 *ibid.*, p. 303

87 GATES, S. P.: 'Notes on the aerodynamics of automatic directional control', *RAE Report No. BA 487*, 19 February 1924. Other attempts at analysis of

aircraft control systems include GATES, S. B.: 'Notes on the aerodynamics of an altitude elevator control', *RAE Report No. BA 494*, 12 March 1924 and GARNER, H. M.: 'Lateral stability with special reference to controlled motion', *ARC R & M 1077*, Oct. 1926

88 GATES: Report 487, p. 7

89 *ibid.*, p. 10

90 *ibid.*, p. 24

91 MELVILL-JONES, B.: 'Dynamics of the aeroplane' *in* DURAND, W. F. (Ed.): *Aerodynamic theory* (Dover, New York, 1963), Vol. 5; quoted from McRUER, D., and GRAHAM, D.: 'A historical perspective for advances in flight control systems', *AGARD Advances on Control Systems*, No. 137, 1974, pp. 2-1–2-7

92 MINORSKY: *op. cit.*, pp. 305–306

Chapter 5

The new technology: electricity

Not long ago, our period was described as 'the century of steam', and rightly so . . . But hardly had the slave *Steam* grown to its full strength when there appeared for the service of mankind a young giantess who as it seems desires to work in harmony with Brother Steam for their masters, but in fact is proceeding completely to displace him. The giantess is *Electricity*. After remaining behind for decades in the process of growth, she suddenly about 25 years ago began to develop; and in this brief time she has made such growth that she begins to revolutionize technology. It is no longer observed that she spent long childhood years in laboratories and that her first steps were guided by quiet scholars.

Arthur Wilke, 1893

Introduction

James Watt provided the means of one revolution through the development of the steam engine; in 1806 the company of Boulton and Watt showed the way to another. For in that year they used gas lighting in their Soho works, the light which 'Andrew Ure claimed adequately replaced sunlight, so that there was no moral obliquity in forcing children to work a twelve-hour day in the factories',[1] which encouraged the habit of reading, which improved the amenities of public places and which brightened the Victorian home. Gas was to remain the major illuminant until well into the 20th century, but as early as 1809 its eventual replacement, electricity, was used to produce light, when at a public lecture at the Royal Institution Sir Humphrey Davy demonstrated an arc lamp. This was the result of just one of the investigations which scientists had been carrying out concerning the nature of electricity, investigations which had been made easier by the development of the voltaic cell in 1800. In September 1831 Michael Faraday showed that the interaction of electric and magnetic fields could produce mechanical motion, and, one year later, the converse, the production of electric current from the interaction of mechanical motion and a magnetic field, was demonstrated by Hippolyte Pixii. By 1834 rotating coil generators were being manufactured commercially in London: the laboratory door was open.

It was many years, however, before arc lamps were used other than for experiment. In 1858 they were installed at the South Foreland lighthouse, but their commercial use did not become widespread until the 1870s. The use of incandescent lamps, electric motors and the development of central generating stations followed soon after.

With the new technology came new control problems: arc lamps required control of the distance between the electrodes, they also required constant current, the incandescent lamp required a constant voltage supply, motors were required to run at constant speed and the development of the alternating-current system demanded frequency control.

Arc lamps

> From Calais pier I saw a brilliant sight,
> And from the sailor at my side besought
> The meaning of that fire, which pierced the night
> With lustre, by the foaming billows caught
> 'Tis the South Foreland!' I resumed my gaze
> With quicker pulse thus on the verge of France,
> To come on England's brightness in advance!
>
> Charles Tennyson Turner, 'The South-Foreland electric light', *Collected sonnets*, 1884

The development of a practical system of light required the solution of three problems: the provision of better material for the electrodes, the development of an economic source of electrical current and the development of a mechanism for regulating the distance between the electrodes so as to produce light of constant intensity.

Arc-length regulation fascinated inventors of the 19th century almost as much as did the problem of engine governing; hundreds of patents were granted, but, as with the governor, few were successful. The control problem is threefold: (*a*) to obtain an arc of constant intensity, the electrode gap has to be maintained constant, but the electrodes are slowly consumed at an irregular rate, (*b*) to relight the lamp the electrodes have to be brought into contact and (*c*) for economic reasons the circuit has to supply several lamps; the control system of any one lamp, therefore, must not act in such a way as to make the overall system unstable.

Most of the early attempts at regulation were concerned with single lamps in which an electromagnet was placed in series with the lamp; as the electrodes burn away the current, and hence the pull exerted by the electromagnet, is reduced. In a patent of 1847, W. E. Staite proposed using the electromagnet to operate a changeover gear; clockwork drove the electrodes together or apart in accordance with the position of the changeover gear, the speed of the movement being regulated by a friction governor.[2] In another patent, issued in the following year, Staite abandoned the clockwork mechanism in favour of a weight to draw the carbon electrodes together and an electromagnet to separate them.[3] In this same year, 1848, Léon Foucault designed an arc lamp which was built by an instrument maker Jules Duboscq and sold under his name with considerable success.[4] The mechanism used was similar to that of Staite's 1847 patent. In another patent of 1848 F. Allman proposed the use of the thermal effects of the current passing through various forms of resistance, as well as the use of a weight and electromagnet.[5]

These early electric arc lights were designed to operate from voltaic cells and were only suitable for use on special occasions. In the mid-1870s the self-excited electric generator provided a cheap source of electrical energy, making it economical to use arc lamps for street lighting, and the illumination of public and other large buildings. In rapid succession a number of technically sound systems of arc lighting emerged. The Gare du Nord in Paris, and mills at Mülhausen and Menier were all lit by arc lamps in 1875; of most interest, however, are the systems which were developed by the Brush Electric Company* and by Elihu Thomson and Edwin J. Houston.†

---

* The Brush Electric Company was formed by Charles F. Brush (1849–1929), who qualified as a mining engineer but between 1877 and 1891 devoted himself to electrical engineering. The company was taken over in 1891 by the General Electric Company [USA], after which Brush turned to science.
† Edwin James Houston (1847–1914) and Elihu Thomson (1853–1937), both graduated from the Central High School, Philadelphia, and both eventually became professors at the School, Houston of physical geography and natural philosophy, Thomson of chemistry and mechanics (which included the responsibility for teaching physics and electricity). In 1878 they co-operated on the design of an arc lamp, and a company was formed to manufacture the equipment. Houston remained at the High School and was not associated with the business after 1882, while Thomson became chief engineer of the Thomson–Houston Company, and on the merger with Edison's company became consultant to the new General Electric Company.

Success went initially to the Brush Company, with the installation in 1878 of 20 arc lights in Wanamaker's Philadelphia department store. This system, patented in 1878, was suitable for use only in single-lamp circuits, but it included a simple friction-clutch arrangement (Fig. 5.1), which was to become standard on all subsequent lamps of the company. The bottom carbon electrode of the lamp is fixed and the upper is attached to a metal rod which passes through the centre of the solenoid. A flat loose washer fits over this rod and is partially supported by a plate attached to the core of the solenoid: with no current flowing the washer is held horizontally and hence the steel rod is allowed to fall, thereby making contact with the lower electrode and establishing the arc; with current flowing the solenoid is energised, it pulls the core upwards, tilting the washer and thereby seizing the rod and drawing it upwards with the core. The upwards force exerted by the solenoid is balanced by the weight of the mechanism and the force which is exerted by the compressed spring.

In this first design the solenoid was in series with the arc (Fig. 5.1b), but in a second design of 1879 an additional coil, (B in Fig. 5.1c) of much higher resistance and wound in the opposite sense was connected across the arc and first solenoid coil. The force on the solenoid coil is now proportional to the difference between the currents flowing in the two coils. Normally only a small fraction of the current will pass through coil B, but if the arc is extinguished the current flowing through the shunt coil will increase and assist in drawing the electrodes together, but, more importantly, it also acts as a bypass, permitting other lamps in the circuit to continue operating.

In 1879 the rival Thomson–Houston system was patented, and its circuit is shown in Fig. 5.2. The shunt coil B controls an electromagnet which when operated withdraws a pawl allowing the upper electrode to descend. The coil A operates against a spring which is attemting to push the lower electrode upwards, reducing the arc gap. The purpose of this solenoid is simply to establish the arc when the current is first applied. The advantage claimed for the Thomson–Houston shunt coil arrangement was that it was unaffected by sudden variations in arc resistance; instability of arc behaviour is a problem which still taxes control engineers today, although not now in connection with arc lamps and currents of a few amperes, but in the control of electric-arc furnaces with currents of several hundred amperes.

The Brush and Thomson–Houston systems were not the first to employ a shunt coil: Werner Siemens in 1873 patented an arc lamp in which a coil in series with the arc operated a ratchet mechanism which separated the electrodes, and a high-resistance shunt winding

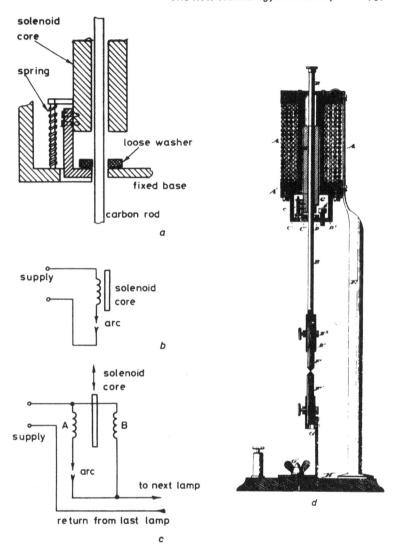

Fig. 5.1 *Brush arc-lamp regulating system*
    a Detail of loose washer, 1878 design
    b Solenoid connection, 1878 design
    c Solenoid connection, 1879 design
    d Drawing from patent of 1878
    [Fig. 5.1*d* is reprinted by permission of the Smithsonian Institution Press from *Feedback mechanisms*, O. Mayr, *Smithsonian studies in history and technology:* number 12: Figure 122. Washington, DC: Smithsonian Institution Press, 1971]

acted to bring the electrodes together.[6] Lantin and Clark in 1877 suggested a mechanism in which the expansion of a rod due to the heating of a shunt coil was to be used to control the separation of the electrodes.[7]

After 1890 interest in the use of arc lamps for general lighting purposes declined, but they continued to be used for military purposes. The First World War resulted in considerable expenditure by the General Electric Company (USA) and, to a lesser extent, by the Sperry Company, on arc-lamp research.

Sperry had become involved in arc lighting in the early 1880s and his company, the Sperry Electric Light, Motor & Car Brake Company, was responsible for numerous small installations throughout the USA between the years 1883 and 1887. These systems were not commercially significant: they did, however, demonstrate a trait which was to characterise the whole of Sperry's work, an instinctive appreciation of feedback control. During this period of his life (1883–1887) Sperry obtained, by his standards, relatively few patents, only 19, but of these 11 included some form of automatic control.[8] The military searchlight shows the ingenuity of the mature Sperry: besides the automatic control of the carbon feed the arc crater was automatically maintained at the focal point of the reflector mirror. The position of the arc was sensed by a differential thermostat which signalled a solenoid to adjust the positive electrode.[9] A duplicate of this searchlight system was used in 1924 by A. A. Michelson in his measurement of the speed of light.[10]

Regulation of current and voltage

The commercial success of the Thomson–Houston lighting system was due largely to its overall current control system, not to the method of electrode position regulation. Experience had shown that arc lamps functioned best at a potential drop of 40 to 50 V and with currents between 6 and 10 A, conditions most easily achieved in the 1880s and 1890s by operating the lamps in series. But arc lamps are very sensitive to variations in line current, and it therefore becomes necessary to have some means of current regulation in the circuit.

One method of regulating the output of a generator is to move the brushes relative to the neutral axis, thereby changing the number of conductors in the armature circuit, and hence the output voltage. A number of early schemes used this technique and, as might be expected, attempted to use a mechanical governor to shift the brushes to compensate for changes in the speed of the prime mover. In 1880 to 1881

Sperry produced a system, using a shaft governor, which relied solely on the mechanical governor. Such systems are, of course, open loop. The Thomson–Houston regulators of 1879 to 1880 used, in addition to a mechanical governor, an electromagnet in series with the output circuit, thus providing closed-loop control (Fig. 5.3).[11] Later Thomson–Houston regulators dispensed with the mechanical governor and used an arrangement shown in Fig. 5.4. As shown in the Figure the solenoid was fitted with an oil dashpot.

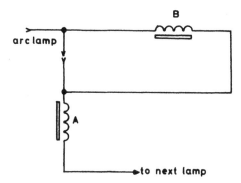

Fig. 5.2   *Thomson–Houston arc-lamp solenoid connection*

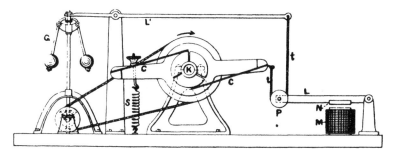

Fig. 5.3   *Thomson–Houston current regulator of 1880*

Brushing shifting tends to give rise to sparking problems and for this reason such systems were quickly abandoned in favour of systems based on modifying the magnetic flux in the machine. Since the magnetic flux depends on both the excitation (number of ampere–turns) and on the reluctance of the magnetic circuit, it can be varied by changing either or both of these quantities. Methods used to change the magnetic circuit included (*a*) moving the pole pieces relative to the armature, (*b*) varying a gap in the magnetic circuit, (*c*) withdrawing the armature from between the poles and (*d*) shunting part of the magnetic field. None met with success except in small machines.[12] The excitation was varied with the aid rheostats and commutators, either by hand, or automatically by means of special governors;

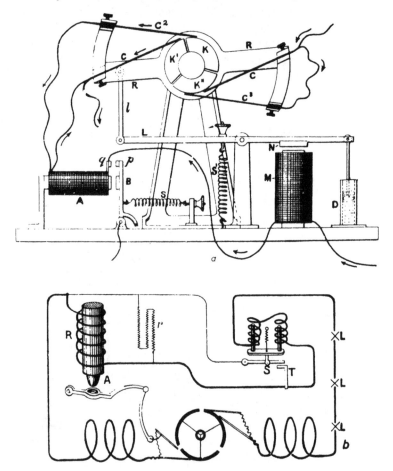

Fig. 5.4   *Thomson–Houston current regulator*
a Regulating gear for Thomson–Houston system
[Reprinted by permission of the Smithsonian Institution Press from *Feedback mechanisms*, O. Mayr, *Smithsonian studies in history and technology:* number 12: Figure 139. Washington, DC: Smithsonian Institution Press, 1971]
b Circuit diagram for Thomson–Houston system
[Thompson, S. P.: *Dynamo electric machinery* (1872), p. 472]

alternatively, compound windings were used to modify the excitation. A typical and widely used excitation regulator was that designed by Woods. His machines were to be found in 'The largest arc-lighting station in the world, that at St. Louis, Missouri . . . with 53 of these dynamos each capable of feeding 60 arc lamps'.[13] The circuit for Wood's regulator is shown in Fig. 5.5. The two main brushes B are fixed and the pilot brush C is moved relative to B by the action of the solenoid. Connected between B and C is a coil wound in the opposite sense to the energising coil; with all the lamps in the circuit

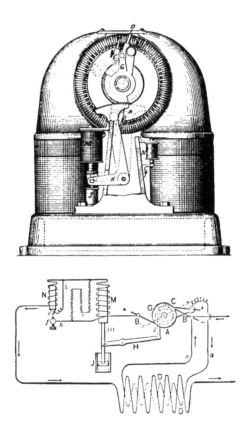

Fig. 5.5 *Wood's dynamo and control circuit*
[Thompson, *op. cit.,* pp. 476, 774]

C is held close to B and hence there is only a small current flowing in circuit *a–a.* If the current in the main circuit increases, C is moved away from B, the feedback or demagnetising current is thereby increased and the excitation is reduced.

An alternative method of varying the excitation, which was used

by Brush, is shown in Fig. 5.6.[14] In this system the change in current in the lamp circuit acting through the solenoid B changed the compression and hence the resistance of the carbon pile C, thereby providing a variable-resistance shunt for the field winding.

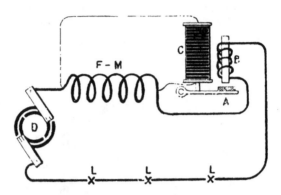

Fig. 5.6   *Brush's automatic current regulator*
[Thompson, *op. cit.* p. 772]

Towards the end of the century, alternating current began to be used for arc lighting; the basic series circuit was retained and hence constant current was still essential. Methods similar to those used for direct-current circuits were proposed, but very rapidly the constant-current transformer was generally adopted.[15]

Arc lamps eventually gave way to the incandescent lamp. The use of carbon filaments in lamps was patented in America by J. W. Starr in 1845, and in England Joseph Swan produced a carbon filament lamp in 1848; a metal filament lamp was demonstrated by Staite in 1848.[16] Thomas Alva Edison (1847–1931) carried out intensive investigations on the best material for the filaments and on the best combination of supply voltage and filament resistance. His conclusions were that the lamps should be operated in parallel with the highest supply voltage consistent with public safety. Operation in parallel, of course, implies operation at constant potential and hence attention was directed towards constant-potential generators.

The methods used to obtain constant current can all be adapted to maintain constant potential by simply changing the detector winding from being series wound and of low resistance, to being shunt wound and of high resistance. However, for constant potential an alternative method is available, the use of 'compounding' to make the machine self regulating.

Compounding was deduced theoretically by Marcel Deprez in 1881 and was almost immediately widely adopted.[17] Deprez's proposal was that a separately excited coil be used in addition to the normal series field winding. Charles Brush had three years earlier used a machine with a compound winding, with the arrangement shown in Fig. 5.7. The field magnets were partly excited by a winding in series with the main circuit and partly by a shunt winding. There is some doubt as to who first invented 'compound winding', as the combination of series and shunt windings became known. Thompson claims that Brush was the first to use it commercially: claims for its invention were also made by Edison, and it was patented in Britain by Crompton and Kapp.[18]

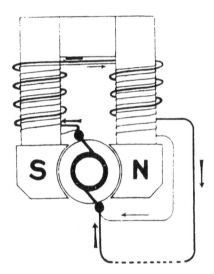

Fig. 5.7   *DC motor with series/shunt winding (Brush, 1878)*
[Thompson, *op. cit.*, p. 59]

The electromagnet, however, continued to be used and it came to be recognised as a general controller similar to the governor. S. P. Thompson described the principle in the following way:

> In all automatic regulators there is part which has to act as the brain of the instrument, watching as it were against any variation, and setting into action the mechanism which is to counteract the variation. This watching device is usually some sort of an electromagnet, often a coil with a movable plunger. When the volts are to be kept constant the coil of the controlling device must be wound as a voltmeter coil, that is of fine wire, of high resistance,

and connected as a shunt. When the amperes are to be kept constant the controlling coil must be wound like an amperemeter with thick wire, of low resistance, and inserted in the main circuit. Alternators are usually regulated by operating on the circuit of their exciters, the current in the governor coil being derived from the mains by a small transformer.[19]

Thompson also advocated using the electromagnet to regulate the steam engines used to drive generators: 'No centrifugal governor attached to the steam-engine can keep the speed of the dynamo truly constant . . . Few mechanical governors will keep the speed within 5 per cent of its proper value, under sudden changes of load. Hence . . . the admission of steam from the boiler to the engine should be controlled by the electric current itself'.[20] In 1881 Richardson devised a governor which could be used either for constant-current or constant-voltage regulation. The measuring and actuating element (Fig. 5.8) was a solenoid acting directly on the throttle valve.[21] P. W. Willans's relay governor of 1883 again used a solenoid, which operated a small valve which controlled admission of fluid to a hydraulic piston.[22] By this means a solenoid consuming only 32 W was able to regulate a 60 h.p. engine.[23] As an alternative to these proposals, dynamometric methods were also suggested. Constant current can be obtained from 'a dynamo driven by a steam-engine governed *not by a centrifugal governor to maintain a constant speed,* but *by a dynamometric governor to maintain a constant torque or turning moment'.*[24] Thompson's final comment on electrical regulation was that 'to render the system truly automatic, it is conceivable that mechanical stoking appliances might be arranged, under the control of the amperemeter or voltmeter, to supply fuel in proportion to the number of lamps alight'.[25]

With the demise of direct-current systems, direct electrical governing of the prime mover went out of use: 'In alternating current work, it is desirable to maintain a given number of cycles, and to govern voltage by electrical means'.[26] A constant number of cycles means a constant speed and a return to the centrifugal governor for frequency control, with the use of electrical regulators for voltage control.

Electrical distribution and power-system stability

. . . electricity's far greater independence of the place where it is generated enables it, however distant from the latter, to carry out its work, while steam can only venture forth a short distance away from its boiler . . . But electricity flies with lightning speed along thin wires over heights, depths and round corners, and is easily distributed, performing its duties in many places simultaneously.

A. Wilke, 1893, quoted from F. Klemm, p. 356

Electrification plus socialism equals communism.
V. I. Lenin

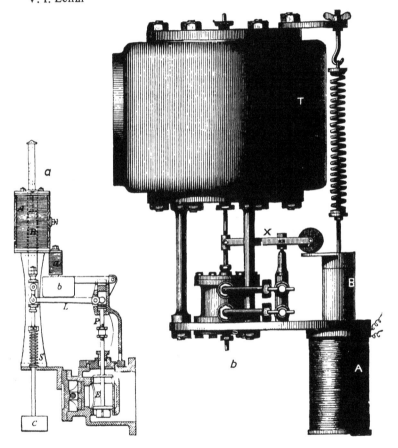

Fig. 5.8 *Electrical steam-engine governors*
  a Richardson's governor, 1881
  [Thompson, *op. cit.*, p. 778]
  b Willan's governor, 1883

At the beginning of the 19th century it had been realised that to make gas available to domestic, commercial and small industrial users it had to be produced centrally and distributed through pipes; by the end of the century extensive distribution networks had been built up.

Similarly, Joseph Bramah had the vision of the distribution of hydraulic power through pipes.[27] Such schemes were implemented in Hull (1877), London (1883), Liverpool (1888), Birmingham (1891), Manchester (1894) and Glasgow (1895).[28] To compete with the established 'gas company' and the developing 'hydraulic company', electricity needed to be generated centrally and distributed to subscribers.

At first electrical distribution was limited to the supply of surplus power from a particular installation, as from the Grosvenor Gallery Company in London in 1883, but these rapidly grew into public supply companies. The benefits of central generation had been recognised earlier by Edison. The first Edison generating station at Pearl Street, New York, was opened in 1882, and by 1886 the Edison system had over 50 stations supplying power for over 150 000 lamps. The Edison system was able to grow rapidly because it used well-established techniques, being a 240 V direct-current system. Edison's continued insistence on the superiority of direct current led to the rapid decline of the Edison Company and in 1892 it was forced to merge with the Thomson–Houston Company to give birth to the General Electric Company.[29]

In the late 1880s George Westinghouse in the USA and Sebastian Z. de Ferranti (1864–1930) in England were demonstrating the superiority of the alternating-current system for electrical distribution. Ferranti was appointed chief engineer to the Grosvenor Gallery Company in 1886 and within two years completely redesigned and re-equipped the station: the distribution voltage was doubled to 2400 V and small transformers were installed in the cellars of subscribers to reduce the voltage to the 100 V then used for domestic purposes. In 1887 a new company, the London Electricity Supply Corporation Ltd, was formed to build and operate a large station proposed by Ferranti. It was designed with sufficient capacity to light the whole of London: power was to be generated at 10 000 V, reduced at local substations to 2400 V for distribution along street mains and then as before reduced to 100 V on the premises of subscribers. Deptford was chosen as the site for this new power station and in October 1889 two Ferranti 1000 kW alternators began supplying power to the system. Four years later the hydroelectric station at Niagara Falls was opened and the era of alternating current had begun.

In these early systems, where the load was largely lighting circuits, the emphasis was on voltage (or current) regulation, the frequency being allowed to vary, under the action of the mechanical governor, with variations in load. With the gradual increase in the proportion of frequency-sensitive components in the load -- for example, the motors used in textile and paper mills – maintenance of frequency became more important. For a short period the increase in parallel operation of generating stations with the consequent increase in system inertia acted as a palliative, but by the 1920s improvements in frequency control were being sought. At the same time the movement towards the interconnection of power systems, with the exchange of power across the lines, was beginning. This brought into focus the question of system stability. F. H. Clough, discussing the stability of large power systems in 1927, paid close attention to regulation, recommending that 'The turbo-alternator has, or should have, two types of automatic governors, one operated by change of speed to control the power of the turbine, and the other operated by change in voltage to control the excitation of the alternator'.[30] He paid particular attention to the speed of response of the regulator and exciter, recommending the use of a vibrating type of regulator which 'responds very quickly to changes of voltage',[31] and in the reply to the discussion he indicated that derivative action was being used to speed up the response: 'use is being made of the transient current induced in the field circuit to operate the regulators even before the voltage of the alternator has actually changed, and thus get the maximum speed of response'.[32]

The most widely used vibrating regulator was the 'Tirill' regulator, which had been designed by the General Electric Company (Schenectady) in 1902. A schematic diagram of the regulator is given in Fig. 5.9*a* and a block diagram in Fig. 5.9*b*. The d.c. control circuit causes the relay to vibrate continuously, and hence the exciter rheostat is continually switched in and out of the circuit. The a.c. control magnet varies the mark/space ratio of this vibration, and, since the exciter time constant is made small with respect to the generator field time constant, the d.c. circuit provides approximately proportional action. Hence the block diagram reduces to that shown in Fig. 5.9*c*.[33]

Many other forms of voltage regulator were developed, by Brown Boveri and Metropolitan Vickers, for example; in the 1930s the General Electric Company developed electronic regulators.[34]

Once attention was turned towards improving frequency control it was quickly realised that 'the *governor alone cannot maintain constant frequency*, and it is necessary therefore to provide some means *outside* of the governor proper to exercise a supervisory control either manually or automatically';[35] a return to the ideas of speed control first proposed by the Siemens brothers. Various frequency standards were developed commercially, the most accurate being the GEC–Warren master frequency regulator based on a compensated 1-second

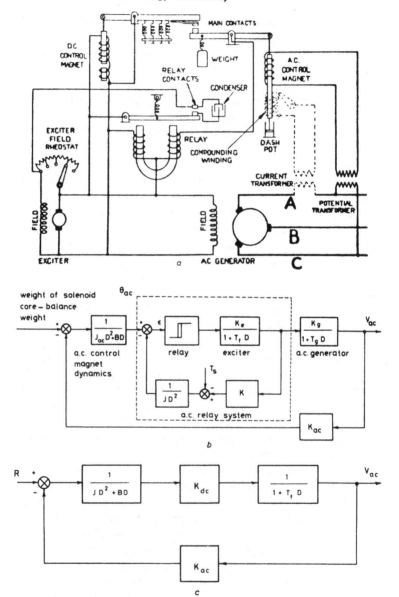

Fig. 5.9 *Tirrill regulator*
  a Schematic
  [Reprinted from Young, H. P.: *Electric power system control* (Chapman & Hall, 1946) p. 38]
  b Block diagram of Tirrill regulator
  c Simplified block diagram

pendulum, with an accuracy of 0·001%.[36] The error signal, the difference between the actual frequency and master frequency, was used to operate a synchronising motor which, as Fig. 5.10a shows, adjusted

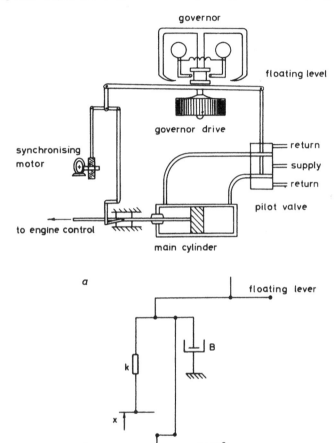

Fig. 5.10   *Frequency compensation*
a Use of synchronising motor to introduce integral-action alternator frequency regulation
b Compensating network used in actual systems

the operating point of the turbine governor. The use of the synchronising motor introduced integral action into the frequency control circuit and hence zero offset. In practice the simple arrangement shown in Fig. 5.10a was not used. A compensating network, as shown in Fig. 5.10b, was interposed between the synchronising motor and the floating lever. The report of the Hydraulic Power Committee shows the arrangements of several other forms of governor with compensating networks, indicating that there was wide recognition of the need to modify the dynamic response of the governor. There is, however, no attempt at a theoretical analysis. A similar lack of dynamic analysis is evident in a group of papers on turbine governors presented at an ASME conference in 1939;[37] during the discussion on one of the papers, R. B. Smith commented that 'Since the question [the governor problem] is essentially dynamic, it seems unfortunate that more analytical and experimental analyses are not common in our technical literature'.[38]

In some ways the lack of analysis of the dynamics of power-system operation is surprising, as the lead had been given at the beginning of the century by the study of the stability of coupled alternators. John Hopkinson[39] had shown in 1884 that a pair of dynamos, mechanically isolated but connected electrically in parallel, tended to keep in step. In practice, however, it was reported by Hopkinson's son Bertram that under certain conditions there was 'a tendency in the machines to develop gradually increasing oscillations about a steady motion in which they are in step with one another'.[40] This behaviour was first investigated by Gisbert Kapp, who attributed it to resonance; Bertram Hopkinson extended the analysis to include the effects of mutual inductance between the field circuit and armature circuit, the self inductance of the armature circuit and the resistance of the armature. He obtained a characteristic equation for the alternating-current motor which was biquadratic, and hence by reference to Routh's *Advanced rigid dynamics* he was able to deduce the conditions of stability.[41] Hopkinson did not use Routh's general criterion for stability, neither did Alexander Russell three years later in his book *A treatise on the theory of alternating currents*.[42] Russell, following Hopkinson's method, obtained a biquadratic equation and says that 'In order to find the required criterion for stability we shall first, by Routh's method, find the products of the pairs of all roots of the biquadratic equation'.[43]

The problem also attracted the attention of Charles Proteus Steinmetz (1865-1923), who showed that the causes of hunting were magnetic lag, pulsation of engine speed, hunting of engine governors and incorrect speed characteristics of engines (to run in parallel the speed of the engines must drop with an increase in load); he suggested the use of damping coils (amortisseurs) to reduce the effects. Throughout the 1920s and 1930s the 2-machine stability problem continued to be studied, since it provided an illustration of the transient behaviour of alternating-current systems and also because many multimachine problems can be reduced to the 2-machine problem. In its simplest form, but without linearisation, the problem is described by a differential equation of the form,

$$\frac{d^2x}{dt^2} + \sin x - C \sin x_0 = 0$$

In the 1930s solutions to this equation were obtained by using mechanical integrators, the differential analyser and step-by-step hand calculator.[44] By using the differential analyser at the Moore School of Engineering, University of Pennsylvania, the stability in the presence of governor deadband was also investigated. In the following year, again using the differential analyser, Concordia investigated the effects of continuous tieline control on the stability of interconnected systems.[45]

The problems of power-system stability although recognised early did not lead to any theoretical developments in control systems, partly because adequate practical solutions were obtained by the use of manual supervisory control,[46] but mainly because the problems were too complicated. Not only were the controllers themselves non-linear, but the problem was multivariable, as was implicitly recognised by Clough in 1927 when he proposed his strategy for regulation.

Electric motors

> One of the inventions most important to a class of highly skilled workmen (engineers) would be a small motive power – ranging perhaps from the force of half a man, to that of two horses, which might commence as well as cease its action at a moment's notice, requires no expense of time for its management and be of moderate price both in original cost and in daily expense.
>
> Charles Babbage, 1851; quoted from D. S. L. Cardwell, *Technology science and history*, 1972, p. 163

Babbage had in mind hydraulic motors, but with the growth of central generating stations it became economical to use small direct-current motors to replace steam or gas engines. Such use led to greater freedom of plant layout and the freeing of overhead space, since the cumbersome line shafting and belt drives were no longer needed. However, for such applications it was 'extremely important that electric motors should be so arranged as to run at a uniform speed, no matter what their load may be'.[47] The centrifugal governor was, of course, used for speed control; it was used to vary rheostats in the armature circuit,[48] the position of the brushes and it was used simply to interrupt the circuit if the motor ran too fast.[49] Ayrton and Perry proposed several regulators in which the current was supplied only for a portion of each revolution, the size of the portion being determined by a centrifugal governor.[50] These methods were all rendered obsolete by development between 1880 and 1884 of compound windings for direct-current motors. A motor with such windings was patented by G. C. André in 1880[51] and the theory of compounded motors was developed in 1883 by Ayrton and Perry.[52] From 1884 onwards numerous motors embodying the principle and showing excellent self regulation were built by F. J. Sprague (1857–1934).[53]

Industry also required motors which as well as having good regulation could be run over a range of speeds. Brush shifting, as well as being undesirable, could only provide a limited speed range, and the use of an armature rheostat gave poor regulation. Various methods of changing the field reluctance were proposed. Edison experimented with a method of varying the amount of metal in the yoke, Diehl and Lincoln produced schemes in which the length of the air gap was to be changed and another method suggested was to make hollow pole pieces in which iron cores could be moved in and out to change the total field flux. None of these methods was particularly practical and attention was soon directed towards variable-voltage systems, characterised by multiple supply voltages, motor–generator systems and 'boost-and-retard' systems. These systems were primarily designed to provide a range of set speeds, control about the set speed being dependent on the inherent speed regulation of the motor.

In was during this period that H. Ward Leonard patented the motor–generator system which now bears his name. The circuit from his patent specification of 1891[54] is shown in Fig. 5.11. He also patented a multiple-voltage system based on the use of three generators,[55] giving six speed settings, and in 1896 he patented a variation on the motor–generator system, the 'boost-and-retard' system, in which the generator was placed in series with the d.c. supply line.[56]

The Ward–Leonard system was used in the operation of gun turrets and platforms on board ships. The so-called Ilgner system, a motor–generator set equipped with a heavy flywheel, became widely used in large rolling mills, the first such installation being a 10 350 hp (13·8 kW) system at the works of Östreichische Berg und Huttenwerke Gessellschaft, Austria, in 1905.[57]

With the increasing dominance of a.c. transmission, by 1900 attention was being turned to the use of induction motors and to methods of obtaining variable speed for these machines. In 1924 Walker listed the methods as:

(*a*) rheostatic control in the rotor circuit
(*b*) resistance and reactance in parallel in the rotor circuit, Boucherot motor
(*c*) change of frequency of supply
(*d*) pole-changing devices
(*e*) cascade connections
(*f*) revaluation of stator
(*g*) the Hunt motor and Creedy modifications
(*h*) a synchronous converter in circuit with rotor (Kraemer's method)
(*i*) a frequency changer in circuit with rotor
(*j*) Leblanc exciter in circuit with rotor, sometimes called the Scherbius method[58]

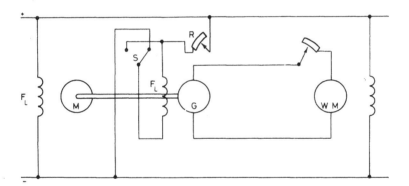

Fig. 5.11   *Ward Leonard motor generator system, 1891*

Rheostatic control in the rotor circuit is wasteful if large speed variations are required, but it was widely adopted for automatic regulation of speed; a typical arrangement is shown in Fig. 5.12. The control motor is supplied through a series transformer in the primary circuit of the main motor; the torque output just balances the weight of the electrodes, but should the current in the main motor increase or decrease the resulting change in torque will cause the electrodes to separate or to move closer together.

The Boucherot technique was used for the United States battleship *New Mexico*, the motors being built by the General Electric Company.[59] Pole changing was also used for warship electric propulsion motors, as was changing the frequency of the supply. The other techniques were widely used in industrial installations. As with the d.c.

machines the emphasis in all applications and analysis was on static characteristics. The control gear was to set the motor to some desired speed; subsequent control of speed depended on the inherent regulation. The 'control gear departments' which grew within the large electrical companies were concerned with the design of contactor mechanisms, starters, control panels, sequencers etc. for motor control and other applications. These were in the main open-loop mechanisms, timers and hand controls, although some starters, for example, used magnetically operated switches, triggered by back e.m.f. or by armature current.[60] Feedback was used in voltage regulation, but this was the concern of the turbo-alternator engineer and the electricity supply industry rather than of the industrial user.

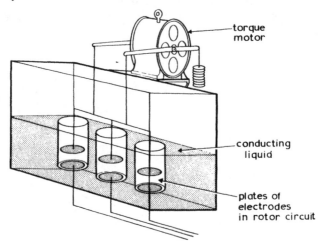

Fig. 5.12   *Liquid slip regulator for a 3-phase motor, c. 1910*
[Reprinted from Sykes, W., *Trans. AIEE*, 1911, **30**, p. 1595]

A. L. Whiteley recalls that when he joined British Thomson Houston in 1930 there were only two areas where feedback control was used:

(*a*) voltage regulators for the field of d.c. exciters for alternators; mainly Tirrel regulators, although rotating contact and carbon pile regulators were also in use
(*b*) sectional drives for paper-making machinery; BTH were using a Schrage motor and controlling speed and power factor by brush shifting, while GEC (Schenectady) were using a motor–generator-set drive for each of the sections. In neither case were stabilising circuits used, the gain was kept low and any hunting was probably masked by the paper-machine disturbances.[61]

Conclusions

The early period of electrical technology saw the development of a number of automatic controls: servomechanisms for controlling arc lamps, voltage and current regulators and methods of motor control. The emphasis was mainly on steady-state behaviour. Controllers were developed by empirical methods and there was little analysis. S. P. Thompson observed that in any regulator there was a part which functioned as a 'brain', but as this 'brain' was typically an electromagnetic device possessing Coulomb friction and other forms of non-linearity analysis was not easy. It was only with the development and growing familiarity with the concept of a time constant that linear approximations of the physical devices began to be used with confidence. Voltage regulators were analysed by Hanna *et al.*[62] and Boice *et al.*[63] in 1939 and 1940, and in 1941 Higgs-Walker[64] in England and Concordia *et al.*[65] in the USA analysed power-system governors.

In all these papers stability was assessed by using the Routh–Hurwitz algebraic criteria. Higgs-Walker obtained the theory from Tolle,[66] while Boice and Concordia make direct reference to Routh's *Advanced rigid dynamics*; Concordia even refers to Maxwell's paper 'On governors'. It is interesting to note that Concordia also expresses the stability of the system by means of stability diagrams in a manner similar to that used by Wischnegradski; an example is shown in Fig. 5.13, where T_1 and T_2 are the governor time constants and the parameter T_d is the damping due to the load.

There was a growing awareness of analysis, of an understanding of theory: Brush, Thomson, Houston, S. P. Thompson and Sprague had all graduated in engineering or science, and Thomson, Houston and S. P. Thomson had studied for a period at German universities. Ferranti, although not a graduate, had studied up to the age of 18 before joining Siemens Brothers and training with William Siemens. They all therefore, combined theory and practice, unlike Edison who was 'for the most part, a trial-and-error inventor. At the time of his search for a lamp filament, he is said to have tried successively some six thousand kinds of vegetable fibers. In most of his work he scorned scientific theory and mathematical study which might have saved him time'.[67] But, even though he scorned scientific theory, Edison recognised that a scientific – a methodical – approach could produce inventions. To this end, in 1876, Edison set up at Menlo Park what was probably the first industrial research laboratory in the world. With the merger, in 1892 of Edison's company and the Thomson–Houston Company to form the General Electric Company, the research laboratory formed the model for the General Electric Research Laboratory, which formally came into being in 1900. The pattern of research and development was changing; the emphasis was moving away from the lone inventor towards the research team.

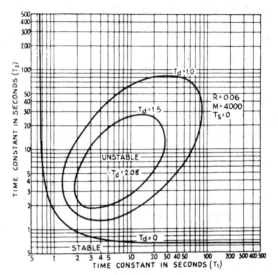

Fig. 5.13 *Stability diagram used by Concordia*
[Reprinted form Concordia, C., *Trans AIEE*, 1941, **60**, p. 559]

Around the same period the Bell Telephone Company was also building up a research team, although formal incorporation of the Bell Research Laboratories did not take place until 1925. These two laboratories and the research and development departments of the British subsidiaries of General Electric, British Thomson–Houston and Metropolitan–Vickers were to play an important part in the development of control engineering.

References and notes

1 DERRY, T. K., and WILLIAMS, T.: *A short history of technology* (Oxford University Press, 1970), p. 513

2 STAITE, W. E.: British Patent 11 783, 3 July 1847

3 STAITE, W. E.: British Patent 12 212, 12 July 1848

4 MAYR, O.: *Feedback mechanisms* (Smithsonian Institute Press, Washington, 1971), p. 105

5 ALLMAN, F.: British Patent 12 276, 28 September 1848, 'Apparatus for the production of light from electricity

6 SIEMENS, W.: British Patent 2006, 5 June 1873; see also SIEMENS, W.: *Scientific and technical papers by Werner von Siemens* (Murray, London, 1895), vol. 2, pp. 341–344

7 LANTIN, Clark: British Patent 2094

8 HUGHES, T. P.: *Elmer Sperry: inventor and engineer* (Johns Hopkins, Baltimore, 1971), p. 45

9 *ibid.*, p. 218

10 *ibid.*, pp. 292–293

11 THOMSON, E., and HOUSTON, E. J.: US Patent 223 659, 20 January 1880; see also MAYR: *Feedback mechanisms*, p. 121

12 THOMPSON, S. P.: *Dynamo-electric machinery: a manual for students of electrotechnics* (E. & F.N. Spon, London, 1892), p. 769

13 *ibid.*, p. 476

14 *ibid.*, p. 772

15 MAYR: *Feedback mechanisms*, p. 123

16 DERRY & WILLIAMS: *op. cit.*, p. 632

17 THOMPSON: *op. cit.*, p. 18, 296

18 *ibid.*, p. 61

19 *ibid.*, p. 769

20 *ibid.*, p. 777

21 RICHARDSON, J.: British Patent 288, 1881; see also THOMPSON: *op. cit.*, p. 778 and RICHARDSON, J.: 'The mechanical and electrical regulation of steam engines', *Minutes & Proceedings of the Institution of Civil Engineers*, **120**, 1885

22 WILLANS, P. W.: British Patents 1184, 5291 and 5945 of 1883

23 THOMPSON: *op. cit.*, p. 779

24 *ibid.*, p. 781 [his italics]

25 *ibid.*, p. 782

26 TRINKS, W.: *Governors and the governing of prime movers* (Constable, London, 1919), p. 147

27 BRAMAH, J.: British Patent 3611, 31 October 1812

28 MacNEIL, I.: 'Hydraulic power transmission: the first 350 years', *Transactions of the Newcomen Society*, 1974–1976, **47**, p. 155; the London system still survives

29 REYNOLDS, T. S., and BERNSTEIN, T.: 'The damnable alternating current', *Proc. IEEE*, 1976, **64**, p. 1343

30 CLOUGH, F. H.: 'Stability of large power systems', *J. IEE*, 1927, **65**, p. 656

31 *ibid.*, p. 657

32 *ibid.*, p. 673

33 Vibrating regulators used with feedback to give proportional action had also been developed for temperature control; for example, the Guoy regulator – see SLIGH, T. M.: 'Guoy regulator', *J. Am. Chem. Soc.*, 1920, **42**, p. 60

34 For details, see YOUNG, H. P.: *Electric power system control* (Chapman & Hall, London, 1942). The subsequent development of electronic voltage stabilisation can be followed by reference to BENSON, F. A.: *Voltage stabilized supplies* (Macdonald, London, 1957) or PATCHETT, G. N.: *Automatic voltage regulators and stabilizers* (Pitman, London, 1970, 3rd edn.), both of whom give extensive bibliographies

35 Hydraulic Power Committee: 'Hydraulic turbine governors and frequency control', *National Electric Light Association Proceedings*, 1931, **88**, p. 544

36 *ibid.,* p. 552; see pp. 566–572 for details of other devices. See also WARREN, H. E.: 'Better frequency control', *General Electric Review*, 1918, **21**, pp. 816–819 for a detailed description of the Warren master frequency controller

37 Annual Meeting, Philadelphia, December 1939; papers published in *Trans. ASME*, 1940, **62**, pp. 167–235

38 SMITH, R. B.: discussion on paper by CAUGHEY, R. J.: 'Steam-turbine governors', *Trans. ASME*, 1940, **62**, p. 197

39 HOPKINSON, J.: 'On the theory of alternating currents, particularly in reference to two alternate-current machines connected to the same circuit', *Journal of the Society of Telegraph Engineers*, 1884, **13**, p. 496

40 HOPKINSON, B.: 'The "hunting" of alternating-current machines', *Proc. Roy. Soc.,* 1904, **72**, p. 235

41 HOPKINSON, B.: *op. cit.,* pp. 237–241. The work of Hopkinson provides more evidence of the close ties between members of scientific and technical circles at the end of the 19th century. Bertram Hopkinson was the son of John Hopkinson, who had been coached by Routh, and in 1871 had emerged as Senior Wrangler and first Smith's prize man; see HILKEN, T. J. N.: *Engineering at Cambridge University 1783–1965* (Cambridge University Press, 1967), p. 99. Bertram himself was placed in the first division of the First Class of the Mathematical Tripos in 1896, and was also connected through marriage to Sir William Siemens. See also MAYR, O.: 'Victorian physicists on speed regulation: an encounter between science and technology', *Notes and Records of the Royal Society of London*, 1971, **26**, pp. 205–228

42 RUSSELL, A.: *A treatise on the theory of alternating currents* (Cambridge University Press, 1906), vol. 2

43 *ibid.,* p. 216

44 CRARY, S. B.: *Power system stability* (Wiley, New York, 1947), vol. 2. Mechanical integrators were used by SUMMERS, I. H., and McCLURE, J. B.: 'Progress in the study of system stability', *Trans. AIEE*, 1930, **49**, pp. 132–158; step-by-step methods were used by BYRD, H. L., and PRITCHARD, S. R.: 'Solution of the two-machine stability problem', *General Electric Review*, 1933, pp. 81–93 and the differential analyser was used by KUEHNI, H. P., and PETERSON, H. A.: 'A new differential analyzer', *Trans. AIEE*, 1944, **63**, pp. 221–228

45 CONCORDIA, C., SHOTT, H. S., and WEYGANDT, C. N.: 'Control of tie-line power swings', *Trans. AIEE*, 1942, **61**, pp. 306–314

46 BOARDMAN, F. D.: 'Control and operation of the UK electricity supply system: an account of its development', *Proc. IEE*, 1978, **125**, pp. 61–65

47 THOMPSON: *op. cit.,* p. 593

48 MAYR: *Feedback mechanisms*, p. 124, for an example of this type of regulator

49 This method was proposed by Brush; see THOMPSON: *op. cit.,* p. 595

50 *ibid.,* p. 595

51 *ibid.,* p. 596

52 AYRTON, W. E., and PERRY, J.: 'Electromotors and their government', *Journal of the Society of Telegraph Engineers*, 1883, **12**, p. 77; see also AYRTON, W. E., and PERRY, J.: 'Note on the governing of electromotors', *Philosophical Magazine*, 1888, **26** (5th series), pp. 63–70

53 THOMPSON: *op. cit.,* p. 596

54 WARD LEONARD, H.: US Patent 463 802, November 1891

55 WARD LEONARD, H.: US Patent 478 344, July 1892

56 WARD LEONARD, H.: US Patent 572 903, 8 December 1896

57 SYKES, W.: 'Electrically driven reversing rolling mills', *Trans. AIEE*, 1911, **30**, Pt. 2, pp. 1606–1616

58 WALKER, M.: *The control of speed and power factor of induction motors* Library Press, London, 1924, p. 48

59 *The Engineer*, 1922, **133**, p. 323

60 EASTWOOD, A. C.: 'Automatic motor controls for direct current motors', *Trans. AIEE*, 1911, **30**, Pt. 2, pp. 1519–1545

61 Private communication

62 HANNA, C. R., OPLINGER, K. A., and VALENTINE, C. E.: 'Recent developments in generator voltage regulation', *Trans. AIEE*, 1939, **58**, pp. 838–844

63 BOICE, W. K., CRARY, S. B., KRON, G., and THOMPSON, L. W.: 'The direct-acting generator voltage regulator', *Trans. AIEE*, 1940, **59**, pp. 149–157

64 HIGGS-WALKER, G. W.: 'Some problems connected with steam turbine governing', *Proc. I Mech. E*, 1941, **146**, pp. 117–125

65 CONCORDIA, C., CRARY, S. B., and PARKER, E. E.: 'Effect of prime-mover speed governor characteristic on power-system frequency variations and tie-line power swings', *Trans. AIEE*, 1941, **60**, pp. 559–567

66 TOLLE, M.: *Die Regelung der Kraftmaschinen* (Springer, Berlin, 1906); Higgs-Walker refers to 3rd edn., 1921

67 *Dictionary of American Biography*, v. XXI, suppl. 1, p. 280

Chapter 6

The shrinking world

Gooch, Heart's Content to Glass, Valentia, 27 July, 6 p.m.: our shore-end has just been laid and a most perfect cable, under God's blessing, has completed telegraphic communication between England and the Continent of America.

The first message sent over the transatlantic telegraph cable, 1866

I then shouted into M [the mouthpiece] the following sentence: 'Mr. Watson — come here — I want to see you.' To my delight he came and declared that he had heard and understood what I said.

Extract from the laboratory notebook of A. G. Bell, reporting the experiment carried out on Friday 10th March 1876

The telegraph and telephone

Communication at a distance has been a long-felt need; smoke signals, drums, trumpets, church bells, beacons and flags are some of the methods which have been used. Telegraphy in the modern sense, however, is a product of the French Revolution. With the French armies fighting on several fronts, rapid communication was essential and Claude Chappe, an ardent supporter of the Revolution, began in 1790 to study the problem. He first considered the use of electricity, but his final recommendation was for the setting up of a series of manual stations equipped with semaphore arms and telescopes to provide rapid long-distance communication. The first line, connecting Paris to Lille, was in operation in 1794 and by the middle of the 19th century the network extended over 4800 km (3000 miles).[1] Under ideal conditions such a system could transmit messages surprisingly quickly, but there were obvious disadvantages: vulnerability to climatic conditions and expense, for example. The future lay with the electric telegraph, the stimulus for which came in the 1830s with the growth of the railways.

In 1837, Wheatstone and Cooke in England and Samuel Morse in the USA gave public demonstrations of the electric telegraph and a year later the electric telegraph was being used by a railway company for signalling between Paddington and West Drayton stations. In 1851 the completion of the Dover–Calais submarine cable linked London to Paris, and in 1866, with the help of the *Great Eastern*, the transatlantic cable was laid, linking the old world with the new.

The growth of the telegraph system was rapid; even more rapid was the growth of the telephone system. Alexander Graham Bell's first public demonstration was in 1876 and by the end of 1877 over 9000 telephones were in use. But problems were not far off. The greater bandwidth required for speech led, as the length of the transmission line increased, to attenuation and distortion. The New York to Chicago line, opened in 1892 over a distance of 1500 km (900 miles), represented the then economic limit of existing technology.

A theoretical investigation of the transmission of signals along telegraph lines had been carried out by Sir William Thomson (Lord Kelvin) as early as 1856. In a highly original paper in which he neglected everything which did not affect the result, including the self induction of the line, he deduced a working principle which became known to telegraph engineers as the 'KR-law' (Thomson used K as the symbol for capacitance).[2] At the time the neglect of the self inductance was justified, since signalling speeds were low. As speeds increased and the simple dots and dashes of telegraphic communications were replaced by the more complex patterns of speech, the simple KR-law ceased to apply. Many engineers were bewildered. Thomas Edison tested his new automatic telegraph apparatus by using a coiled cable 3500 km (2200 miles) long and was astonished when 'a Morse dot normally a thirty-secondth of an inch long was extended into a line about thirty feet long'.[3] Discoveries such as this led the telegraph engineers into frantic efforts to reduce the self inductance of telegraph lines in order, as they thought, to increase signalling speeds and reduce distortion:

> Self-induction's 'in the air'
> Everywhere, everywhere,
> Waves are running to and fro,
> There they are, there they go.
> Try to stop 'em if you can
> You British Engineering man.

wrote Oliver Heaviside (1850–1925) in one of his notebooks.[4] Heaviside devoted most of his working life to the problems of electrical signal transmission, and among his many contribufions to the subject was his demonstration (theoretical) that the *addition* of inductance to telegraph lines could increase signalling speed and reduce distortion. The methods he used were obscure to the engineers of the time: the formulas were complex and the idea was totally opposed to the prevailing notions. Considerable advocacy was required to gain acceptance for the idea and it was not put into practice by Heaviside himself but by two American engineers.

In 1889 Professor M. I. Pupin (1868–1931) and G. A. Campbell (1870–1954) both filed patents for inductive-loading devices. A patent interference hearing granted priority to Pupin, who also published papers in 1899 and 1900 clearly demonstrating the improvements which could be obtained.[5] The American Bell Telephone Company immediately acquired Pupin's patents and the practical development of inductive loading was superintended by Campbell.

George Ashley Campbell had been recruited to the fledgling research laboratories of the American Bell Telephone Company in 1897 after studying at MIT, Harvard, Paris, Vienna and Göttingen. His task was to consider how the recently developed electrical theories could be applied to the improvement of telephone transmission circuits. The fundamental theoretical work covering the principles of loading telephone circuits was completed in 1899 and in September of that year laboratory tests confirmed the theory. In May 1900 inductively loaded lines were put into commercial service.[6]

The use of inductively loaded lines reduced distortion, but the Bell Company's aim of producing a nation-wide network was still frustrated by signal attenuation. Even with very heavy open wire lines it was not possible to transmit voice frequencies across the continent without the use of repeaters.[7] In the absence of electronic amplifiers – the thermionic valve had yet to be invented – they turned to the use of electromechanical amplifiers. Receiver–microphone assemblers, one for each direction, were installed at intervals along the lines. It was rapidly discovered that if these repeaters were incorrectly positioned 'howling' could occur,[8] but there seems to have been little investigation of the phenomenon until the work of Campbell in 1912.

By this time the Bell Telephone Company had standardised on two types of repeater: type 21, in which a single repeating element was used to amplify messages reaching it from both directions, and type 22, which used two amplifying elements, one assigned to each direction of transmission. The type-21 repeater required the two sections of line to be balanced, whereas the type 22 used a network to balance the incoming line (the basic circuit for a type-22 repeater is shown in Fig. 6.1). For type-22 amplifiers,

> singing will not be introduced by any possible unbalance, however large, in either of the lines, provided the unbalance of the other line does not exceed a certain critical magnitude. Furthermore, the two lines connected together may differ radically in character since each is balanced separately against its own artificial line ... Theoretically, a given total amplification can be secured with a larger singing margin if it is distributed among a number of properly spaced points along the line rather than concentrated at a single point

wrote Campbell in 1912.[9]

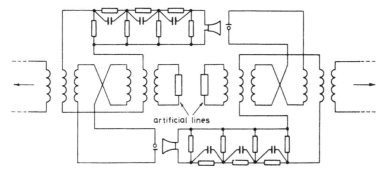

Fig. 6.1   *Type-22 2-way telephone repeater*

While the reduction in gain required for an individual repeater had a beneficial effect on stability, it increased the distortion, since for a given distance more repeaters were required. It thus made increasing demands on the fidelity of the repeater. The problems were compounded by the introduction, in 1918, of a carrier telegraph system, and in 1923 with the voice-frequency carrier system. The higher frequencies of the carrier systems resulted in increased attentuation and hence in a need for a larger number of repeaters. The solution to the repeater problem was to be found in the feedback amplifier, the foundations for the development of which were laid by the pioneers of radio communication.

Radio transmission and positive feedback

From Our Correspondent,

St. Johns, N.F., Dec. 14 – Signor Marconi authorizes me to announce that he received on Wednesday and Thursday electrical signals at his experimental station here from the station at Poldhu, Cornwall, thus solving the problem of telegraphing across the Atlantic without wire. He has informed the Governor, Sir Cavendish Boyle, requesting him to apprise the British Cabinet of the discovery, the importance of which it is impossible to overvalue.

The Times, December 16, 1901

One morning in July 1896 a young man called at the GPO West building St. Martin's-le-Grand, London, with a letter of recommendation to Sir William Preece from A. A. Campbell Swinton. The young man was Guglielmo Marconi; he had with him two large bags containing his wireless-telegraphy apparatus, the operation of which he demonstrated to Preece.[10] For his first experiments Marconi had used as a detector a 'coherer' tube, a glass tube filled with metal filings, which he had to tap to 'decoher' after the receipt of the signal. The Post Office work-shop modified the system so that on receipt of a signal the tube was automatically 'decohered' by being tapped by the hammer of a modified electric bell which was connected to the receiving circuit. It was this form of apparatus, much refined and improved, which was used to transmit the signal s the 2700 km (1700 miles) from Poldhu in Cornwall to Newfoundland on 12th December 1901.

Supervising the operations at the transmitting station at Poldhu was John Ambrose Fleming (1849–1945), a consultant to Marconi's company and formerly a consultant to the Edison Electric Light Company. Fleming was becoming slightly deaf and was therefore beginning to look for a means of visually detecting the receipt of radio signals. He remembered some experiments which both he and Thomas Edison had carried out in the 1880s. The early electric 'glow lamps' had been plagued by deposits forming on the glass, by blackening of the bulb. Both Fleming and Edison agreed that the blackening was due to carbon thrown off by the filament. Edison tried to drive the carbon particles back to the filament by applying an electrostatic charge to a coating of tin deposited on the inside of the glass bulb at the point of maximum discoloration. On connecting the tin foil, via a galvanometer, to the positive side of the d.c. filament supply, a small current was seen to flow, but no current flowed when the tin foil was connected to the negative side of the supply. Fleming duplicated the experiment, but neither he nor Edison carried the investigation any further at that time. In 1901, however, Fleming tried using the 2-electrode lamp to detect the receipt of radiowaves. The circuit he used is shown in Fig. 6.2. In 1904, he patented the thermionic diode.[11]

Two years later Lee de Forest (1873–1961) added a third filament, the grid filament, producing the triode valve, or, as de Forest called it, the Audion.[12] The discovery was announced in a paper published in the *Transactions of the American Institute of Electrical Engineers*[13] and de Forest was granted a patent for it in 1908. The audion appeared to offer a way of providing the much needed increase in the sensitivity of radio receivers, but at first its behaviour was not very clearly understood and considerable effort was expended in investigating its properties. Investigations carried out by Harold D. Arnold[14] in the laboratories of the Western Electric Company and by Irving Langmuir[15] at the Research Laboratories of the General Electric Company showed that if a high vacuum could be obtained in the tube the valve would have a longer life and a more predictable performance.

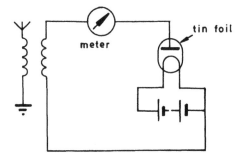

Fig. 6.2   *Fleming's basic radio signal detector, 1901*

The work at the General Electric laboratories had been instigated by E. F. W. Alexanderson[16] late in 1912. Alexanderson had been working on the development of a radio-frequency oscillator for R. A. Fessenden when, early in 1912, in trying to adopt the radio alternator for use as an amplifier for telephonic signals, he invented the so-called magnetic amplifier. That year a wealthy and enthusiastic inventor, John Hays Hammond, purchased two radio alternators for his experiments on radio-controlled torpedo boats. As a result of this purchase Alexanderson visited Hammond's laboratory and was shown the 'ion-controller' (audion) which one of Hammond's assistants was using in some experiments to find a suitable amplifier for radio signals. Alexanderson, interested in the development of a selective and sensitive radio receiver, pushed the GERL into the rapid development of an improved audion and on the 14th May 1913 with Langmuir he witnessed the testing of the new valve. A few days later, a receiver with a 2-stage amplifier was demonstrated.[17]

In the summer of 1912 Edwin Armstrong (1890–1964), an undergraduate at Columbia University, discovered that the gain of the audion could be increased if part of the output signal was *coupled* back to the input circuit. By the autumn of 1912 he had constructed an audion receiver of extreme sensitivity. He had insufficient funds to file a patent, but on 31st January 1913 he had a drawing of the circuit witnessed by a notary and eventually filed a patent application in October 1913. The circuit given in the patent specification showed only audio-frequency feedback, but the accompanying description indicated that Armstrong was also aware of the effect of using radio-frequency feedback. In claim 9 he wrote 'An audion wireless receiving system having a wing circuit [anode circuit] interlinked with a resonant grid circuit upon which the received oscillations are impressed, and an inductance through which the current in the wing circuit flows, the grid circuit including connections for making effective upon that circuit the potential variations resulting from a change of current in the wing circuit' and, in claim 15, 'An audion wireless receiving system having a wing circuit interlinked with a resonant grid circuit upon which the received oscillations are impressed, and means supplementing the coupling of the audion to facilitate transfer of energy from the wing circuit to the grid circuit, whereby the effect upon the grid of high frequency pulsations in the wing circuit is increased'.[18] In a paper published in 1915,[19] Armstrong shows a circuit (Fig. 6.3) incorporating both types of feedback, describing the operation of the circuit as follows:

> Here M_2 represents the coupling for the radio frequencies and the coils are of relatively small inductance. M_3 is the coupling for the audio frequencies, and the transformer is made up of coils having an inductance of the order of a henry or more. The condensers C_3 and C_4 have the double purpose of tuning M_3 to the audio frequency, and of by-passing the radio frequencies. The total amplification of weak signals by this combination is about 100 times, with the ordinary bulb. On stronger signals, the amplification becomes smaller as the limit of the audion's response is reached.[20]

Armstrong was also aware that his audion amplifier could become unstable. It 'is made more stable and shows less tendency to become a high frequency generator and to set up oscillations in the interlinked circuit, if the tuned grid circuit is grounded',[21] but he seemed to be unaware that positive feedback also provided sharper tuning; his emphasis was on the increase in amplification.

At this time, the concept of positive feedback was, as it were, in the air; it was ready to be discovered, and between 1912 and 1914 several people besides Armstrong sensed the concept and in various ways attempted to make use of it. On 6th August 1912 von Etten,

Lee de Forest's assistant, had accidentally discovered that if the output of a double audion was connected to the input, the audion would howl or sing. Neither von Etten nor de Forest seemed to have recognised the importance of this effect or commented further. Slightly earlier than this, in the winter of 1911–1912, there is evidence to suggest that Fritz Lowenstein in the USA had a working audion oscillator, but Lowenstein himself has left no record of this device.[22] Also working on the design of oscillator circuits was Alexander Meissner in Berlin. In 1913 using a Leiben–Reisz valve, a triode valve like the audion, he deliberately introduced positive feedback. In his British and United States patents, filed in 1914, he discusses in detail the application of positive feedback in oscillators, amplifiers, amplifier–detectors and herterodyne receivers.

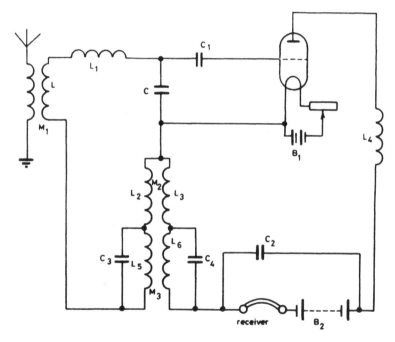

Fig. 6.3   *Armstrong's circuit for combined radiofrequency and audiofrequency feedback, 1915*

In England, Charles S. Franklin of Marconi's Wireless Telegraph Company filed a patent on 12th June 1913 in which he described a regenerative amplifier (Fig. 6.4). He wrote [the italics are his], 'we make the circuit, in which the magnified oscillations occur, *react* on the circuit, in which the oscillations to be magnified occur, by *coupling* these circuits, either electrostatically or electromagnetically, to a certain degree.

If the coupling be too strong the tube will be unstable and will itself tend to produce oscillations but there is a certain critical strength of coupling below which the tube is unable to maintain oscillations. At a coupling a little below this critical strength the tube and circuits are stable but act while receiving oscillations as though the resistance in the circuits was very small.

The result is that the damping of the receiving system can be reduced to any required degree and the tuning of the system is made very sharp.[23]

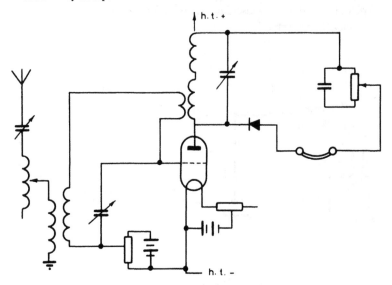

Fig. 6.4   *Circuit diagram of Franklin's regenerative amplifier, 1913*

The above is the earliest clear description of the effect of positive feedback; it is, however, not clear how much debt Franklin owes to Meissner. It is known that he visited Meissner in Germany in the early part of 1913, but it appears that at that time Meissner was concerned only with the generation of oscillations. As Tucker has pointed out, Franklin was the only one of the inventors of the period who mentioned 'the effect of feedback in reducing the damping and sharpening the tuning . . . an important property of the system . . . a benefit in making the receiver more selective, and a nuisance in making it difficult to maintain a receiver in tune for a long period'.[24]

In the receivers of Armstrong and of Franklin the audion was used to provide amplification; positive feedback was used to increase the gain, but the amount was restricted to a level below that at which self oscillation would occur. An alternative method of reception was to use a heterodyne receiver. Two types were developed: the autodyne used for radio telegraphy and the homodyne which could be used for receiving speech transmissions. Both systems required stable oscillators, and suitable circuits, using positive feedback, were developed by Armstrong, Meissner, de Forest, Logwood and Kendall.[25] The increasing use of oscillating circuits led to theoretical studies being undertaken; for example, by Hazeltine.[26] The importance of oscillating circuits was also recognised and studied by van der Pol[27] and E. V. Appleton.[28]

The telephone and the negative feedback amplifier

> The engineer who embarks upon the design of a feedback amplifier must be a creature of mixed emotions. On the one hand, he can rejoice in the improvements in the characteristics of the structure which feedback promises to secure him. On the other hand, he knows that unless he can finally adjust the phase and attenuation characteristics around the feedback loop so the amplifier will not spontaneously burst into uncontrollable singing none of these advantages can actually be realised.
>
> H. W. Bode, 1940

By using heavily loaded lines and electromechanical repeaters a telephone link had, in 1913, been established between New York and Salt Lake City, a distance of 4200 km (2600 miles). It was, however, far from satisfactory: 'nothing worked well enough until the vacuum tube amplifier was invented. This infant device was grabbed enthusiastically, and with its help the continent was finally spanned from New York to San Francisco in 1915'.[29] The conductors used for this first transcontinental link weighed almost 300 kg/km (0·5 ton/mile) and even with a cut-off frequency of 1000 Hz the loss over the circuit was 60 dB. Six repeater amplifiers were used with a total gain of 42 dB, giving a net loss over the circuit of 18 dB.

The path forward seemed straight and broad: vacuum-tube amplifiers, made in large numbers, would be cheap in relation to copper cost so that, by reducing the size of the cable and increasing the number of repeaters, costs could be reduced. Campbell's invention of the wave filter in 1917[30] made it possible to use the lines for more than one conversation at once by using carrier techniques. Long-distance telephoning began to grow rapidly and in 1923 a second transcontinental link via Texas to Los Angeles was established. It used 20 amplifiers and could carry 4 conversations simultaneously. But the path was not smooth: each repeater was a source of noise and distortion and as the number of repeaters grew the greater became the overall degradation of the signal.

Various attempts were made to improve amplifier design; careful selection of the valves, use of matched valves in push–pull arrangements, and the use of multistage amplifiers were all tried. Work on multistage amplifiers led to the discovery that because of feedback through the grid–plate capacity of the valve the amplifier could 'sing' when used with radio-frequency signals. This effect was investigated by Friis and Jensen,[31] who reported in 1924 that for a multistage amplifier the overall gain is given by the product of the ordinary amplification K and the 'feedback' amplification K', where

$$K' = \frac{R}{R - R'} \qquad (6.1)$$

R is the actual resistance of the loop and R' is the equivalent series resistance introduced into the circuit by the 'feedback' action. They noted that R' could be positive, giving an increase in amplification, or negative, giving a decrease, but they investigated only the behaviour for R' positive.

In 1923, a graduate in electrical engineering from Worcester Polytechnic, Harold S. Black (*b.* 1898), joined the laboratories of the Western Electric Company and began working on the design of repeater amplifiers, work which continued when, in 1925, he joined the staff of the Bell Telephone Laboratories. Black's first attempts to reduce distortion in repeater amplifiers were made around 1925 and led to an arrangement similar to that shown in Fig. 6.5. The boxes labelled + represent biconjugate devices similar to three winding transformers or Wheatstone bridges; transmission can take place from any input to either adjacent output, but not to the facing output.[32] The behaviour of this circuit has been described by Bode:

> The top amplifier is the principal one . . . the second amplifier is a compensating device . . . the output signal from the principal amplifier is first reduced by the variable attenuator labelled θ and balanced against a sample from the input circuit. If we set $\theta = 1/\mu$, so that the loss (and phase) of the variable attenuator exactly equals the gain of the top amplifier, and if the top amplifier is perfect in the sense that its output contains no extraneous noise or modulation components, this balance is precise and nothing is fed to the bottom amplifier. However, extraneous noise or modulation from the top amplifier is not balanced out.

Instead, it goes through the bottom amplifier and is combined, with reverse polarity, with the direct feed from the first amplifier ... Since the gain in the second amplifier is just sufficient to balance the loss in the $1/\mu$ attentuator, these imperfections in the original amplifier output are cancelled ... and do not appear in the final output line.[33]

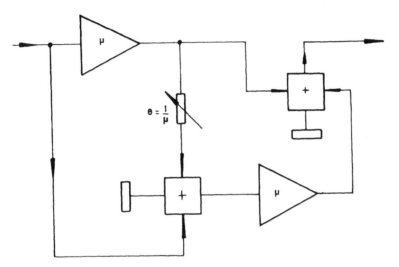

Fig. 6.5   *Black's repeater amplifier, 1925*

Since the compensating amplifier carries only small signals its linearity is not critical, but by making it identical to the main amplifier it can provide a transmission path in the event of failure of the main amplifier, thus increasing the reliability. Experimentally, by careful balancing and tuning, it was possible over a restricted frequency to realise the expected advantages of this circuit; in practice it proved to be difficult to maintain the balances in the biconjugate circuits and the balance between the gain of the amplifier and the loss in the attentuator over a broad frequency band. Black was forced to search for another approach.

The invention arising from this second approach was patented in 1927,[34] but details of the amplifier were not published by Black until 1934.[35] It is in the introduction to this paper that Black gives the now famous description of the method of obtaining a high-quality, stable, linear amplifier:

However, by building an amplifier whose gain is deliberately made, say 40 decibels higher than necessary (10 000 fold excess on energy basis), and then feeding the output back on the input in such a way as to throw away the excess gain, it has been found possible to effect extraordinary improvement in constancy of amplification and freedom from non-linearity. By employing this feedback principle, amplifiers have been built and used whose gain varied less than 0·01 dB with a change in plate voltage from 240 to 260 V and whose modulation products were 75 dB below the signal output at full load ... Stabilized feedback possesses other advantages including reduced delay and delay distortion, reduced noise disturbance from the power supply circuits and various other features best appreciated by practical designers of amplifiers.[36]

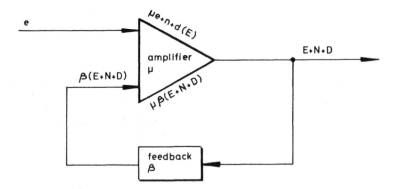

Fig. 6.6   *Black's stabilised feedback amplifier, 1934*
[Redrawn from Black, H. S., Bell Syst. Tech. J., 1934. Reprinted with permission from the *Bell System Technical Journal*, copyright 1934, The American Telephone and Telegraph Company]

e = input voltage
μ = propagation [gain] of amplifier circuit
μe = signal input voltage without feedback
n = noise output voltage without feedback
$d(E)$ = distortion output voltage without feedback
β = propagation of feedback circuit
E = signal output
N = noise output (total)
D = distortion output (total)

Black explained the behaviour of the feedback amplifier with the aid of the diagram shown in Fig. 6.6. He draws attention to the fact that μ and β are complex quantities (functions of frequency) and that μ represents 'the complex ratio of the output to the input voltage of the amplifier circuit',[37] not just the gain of a particular tube. A general equation

$$E + N + D = \frac{\mu e}{1 - \mu\beta} + \frac{n}{1 - \mu\beta} + \frac{d(E)}{1 - \mu\beta} \qquad (6.2)$$

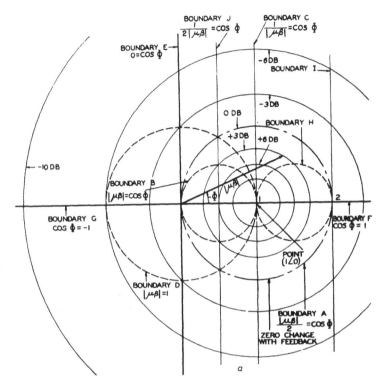

Fig. 6.7   *Design charts from Black's 1934 paper [part b is on p. 194]*
[Reprinted with permission from the *Bell System Technical Journal*,
copyright 1934, The American Telephone and Telegraph Company]

relating signal, noise and distortion is obtained and Black notes that assuming $|1 - \mu\beta| > 1$, in the output, signal, noise and modulation are all reduced. He furthermore says that 'If $|\mu\beta| \gg 1$, $E \doteq - e/\beta$. Under this condition the amplification is independent of μ but does depend upon β. Consequently the overall characteristic will be controlled by the feedback circuit which may include equalizers or other corrective networks'.[38]

Black defines the change in closed loop gain G_{CF} of the amplifier as

$$G_{CF} = 20 \log_{10} \left| \frac{1}{1 - \mu\beta} \right| \qquad (6.3)$$

and then shows that on the vector field of $\mu\beta = |\mu\beta|\phi$, the closed-loop gain is represented by a set of concentric circles about the $(1, 0)$ point;

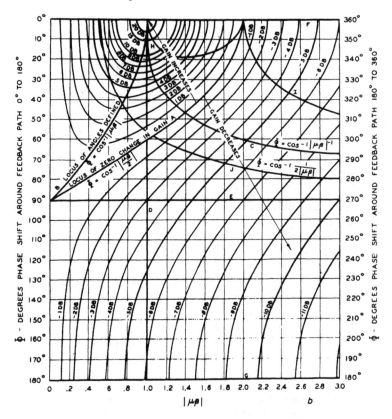

Fig. 6.7*b*

that is, in modern terms, *M*-circles on the inverse polar plot. He also gives the corresponding rectangular plot (phase/attentuation) in a form similar to the Nichols chart. These charts (Fig. 6.7), as Black noted, enable the designed to obtain 'all of the essential properties of feedback action . . . such as change in amplification, effect on linearity, change in stability due to variations in various parts of the system, reduction of noise, etc.'[39]

In considering the benefits accruing from the use of negative feedback Black did not forget the problems: 'It is far from a simple proposition to employ feedback in this way because of the very special control required of phase shifts in the amplifier and feedback circuits, not only throughout the useful frequency band but also for a wide range of frequencies above and below this band. Unless these relations are maintained, singing will occur, usually at frequencies outside the useful range'.[40] Black was, however, puzzled. He had thought that 'singing would result whenever the gain around the closed loop equalled or exceeded the loss and simultaneously the phase shift was zero'[41] (Black was including the phase shift of 180° produced by the valve; plotting of $-\mu\beta$, and hence automatically including the phase shift of the valve was introduced later by H. W. Bode[42]).

In 1927 Harry Nyquist, an engineer working for the American Telephone & Telegraph Company, was asked to assist with the investigation of the conditions under which a feedback amplifier is stable. The result of Nyquist's involvement in this problem was the publication in 1932 of his famous paper 'Regeneration theory'.[43] It was in this paper that, for a restricted case, the stability criterion which now bears his name was first presented. Consideration of the paper itself falls outside the scope of this book; however, the background to the paper, and to the work of Hazen[44] and Ivanoff[45] during the same period, is covered in the following Section.

Development of the operational calculus and its application to electrical circuits

There was a major division between the communications engineer and other electrical engineers. With some minor exceptions, for example the 'hunting' of coupled alternators, electrical engineers were interested in steady-state analysis, and for this purpose the vectorial methods developed by Arthur Edwin Kennelly (1861–1939)[46] and C. P. Steinmetz were appropriate.[47] The communications engineer, however, faced a different problem. His circuits had to follow and reproduce a randomly varying input signal wave with high fidelity; in a sense, his circuits are always in a transient condition. William Thomson (Lord Kelvin) investigated transient behaviour in telegraph cables in 1856 and, as we have already seen, deduced his famous 'KR-law';[48] however, the major development in this field was as a result of Heaviside's work.

Heaviside, completely self taught, had an instinctive understanding of the behaviour of electrical circuits; he realised that in dealing with the transmission problem what had to be considered was not the behaviour of all of the complex wave, but the behaviour of a single pulse, a solitary wave. And what he wished to do was to 'ride on the front of this wave and to study what happened when it met obstacles'.[49]

To 'ride on the front of the wave' Heaviside needed to find a mathematical description of a pulse, which led him to define his unit function $H(t)$. By applying the unit function and the operational calculus, the behaviour of the electrical current in a large number of circuits could be determined. Heaviside in investigating, say, the rate of rise of current in an inductive circuit would have written

$$RC + L\frac{dC}{dt} = E \tag{6.4}$$

where R = resistance, C = current,* L = inductance and E = applied voltage. Then, following Boole, he would have replaced d/dt by the symbol p, to give

$$(R + Lp)C = E \tag{6.5}$$

and would have simplified further by writing a for R/L and C_0 for E/R, giving

$$(1 + p/a)C = C_0 \tag{6.6}$$

In the next step, much bolder than Boole, he would have written

$$C = \frac{1}{1 + (p/a)}C_0 = \frac{a}{p}\frac{1}{1 + (a/p)}C_0 \tag{6.7}$$

which gives

$$C = \left(\frac{a}{p} - \frac{a^2}{p^2} + \frac{a^3}{p^3} - \ldots\right)C_0 \tag{6.8}$$

He would then have applied his rule $1/p^n = t^n/n!$, giving

$$C = C_0\left(a\frac{t}{1!} - a^2\frac{t^2}{2!} + a^3\frac{t^3}{3!} - \ldots\right) \tag{6.9}$$

hence

$$C = C_0\,1 - e^{-R/Lt}, \text{ replacing } a \text{ by } R/L \tag{6.10}$$

Heaviside published an account of his methods in the *Philosophical Magazine* in 1888 and in *The Electrician* over the years 1892 to 1898. He was attacked by the contemporary mathematicians for his lack of rigour; the theoretical foundations for his procedure are in fact very shaky. Expansion of the operator gives rise to an infinite series of terms involving p; $H(t)$ likewise is an infinite series and in the general

---

* Heaviside paid close attention to the symbols used to represent various quantities and was an advocate of the use of C for current and K for capacitance.

$$\frac{1}{\ } = \int_0^\infty A(t) e^{-pt} \, dt \tag{6.16}$$

case E will be an arbitrary function which has to be represented by a Fourier series having an infinite number of terms. Therefore, in general, the use of Heaviside's procedure involves the product of three infinite series. He bitterly resented the attacks, but his only justification was that the rules had been developed on the basis of known cases and that in extending the procedure to other problems knowledge of the physical behaviour of the system must also be considered.

In fact the basis of the method used by Heaviside is implicit in the work of Cauchy, D. F. Gregory, Robert Carmichael and George Boole and is based on the Laplace and Fourier transforms.[50] To what extent Heaviside was aware of this work is difficult to assess. He certainly makes no reference to it, but as a self-taught man, writing at a time when acknowledgment of earlier work was not always made to the extent which would be considered normal today, this is by no means conclusive evidence that he was unaware of the earlier work. However, his contemporaries treated him as the inventor of the operational calculus and several mathematicians at the beginning of this century attempted to justify his methods without reference to the earlier work, thus suggesting that it was not widely known. What is beyond doubt is Heaviside's achievement in the application of the operator technique to a wide range of electrical-engineering problems.

The problems of transient behaviour which Heaviside tackled give rise to equations of the form:

$$f(x) = a_0 \frac{d^n x}{dt^n} + a_1 \frac{d^{n-1} x}{dt^{n-1}} + \ldots + a_n x = \begin{cases} 1 & t > 0 \\ 0 & t < 0 \end{cases} \quad (6.11)$$

and using the Heaviside–Cauchy operational calculus the solution can be written as

$$x(t) = \frac{1}{\phi(p)} H(t) \quad (6.12)$$

where $\phi(p) = a_0 p^n + a_1 p^{n-1} + \ldots + a_{n-1} p + a_n$. As we have seen, Heaviside's procedure was, at this point, to expand $\phi(p)$ in terms of partial fractions. Assuming, for simplicity, that $\phi(p) = 0$ has non-multiple real roots,

$$\phi(p) = \frac{K_1}{p - p_1} + \frac{K_2}{p - p_2} + \ldots + \frac{K_n}{p - p_n} \quad (6.13)$$

Each of the terms $K_i/(p - p_i)$ can then be expressed as a power series in p and Heaviside's 'rule' $1/p^n = t^n/n!$ can be applied.

In 1916, T. J. I'A. Bromwich (1875–1930)[51] recognised that the

solution to eqn. 6.11 was given by

$$x(t) = \frac{1}{2\pi i} \int_{\gamma - i\infty}^{\gamma + i\infty} \frac{e^{\lambda t}\, dt}{\lambda \phi(\lambda)} \tag{6.14}$$

where the path of integration, the so-called Bromwich path, is the line $\mathrm{Re}\,(\lambda) = \gamma$ in the λ plane, γ being real and positive such that all the roots of $\phi(\lambda) = 0$ lie to the left of the line $\lambda = \gamma$.

It can be easily seen that the path of integration for the integral in eqn. 6.14 can be changed to any circle with the origin as the centre and with the zeroes of $\phi(\lambda)$ inside its circumference. The Bromwich form is also seen to be the s-multiplied form of the Laplace transform, the multiplication factor arising because Bromwich was considering the step response.[52] Carson in 1917[53] and Fry in 1919[54] both demonstrated, using Fourier's integral theorem, the validity of Heaviside's expansion theorem for evaluating the Bromwich integral for particular forms of $\phi(p)$.

It had been recognised that the operational impedance $\phi(p)$ was directly related to the steady-state a.c. impedance, the steady-state impedance being obtained by substituting $j\omega$ for p in $\phi(p)$. Then, in 1919, Carson[55] showed that the operational impedance was related to the so-called indicial admittance by the integral equation

$$\frac{1}{Z(p)} = p \int_0^\infty e^{-pt} A(t)\, dt \tag{6.15}$$

Thus the Bromwich integral (eqn. 6.14) and the Carson integral (eqn. 6.15) form a transform pair.

Implicit in Heaviside's approach, but never exploited by him, was a method of determining the response of a network to any arbitrary signal; for as Carson wrote in 1925: 'The indicial admittance of an electrical network completely determines, within a single integration, the behaviour of the network to all types of applied electromotive forces. As a corollary, a knowledge of the indicial admittance is the sole information necessary to completely predict the performance and characteristics of the system including the steady state'.[56] He then goes on to say that

we have reduced the determination of the current in a network in response to an electromotive force $E(t)$, impressed on the network at reference time $t = 0$, to the mathematical solution of two equations: first the integral equation

$$\frac{1}{pZ(p)} = \int_0^\infty A(t)e^{-pt}\, dt \tag{6.16}$$

and second the definite integral

$$I(t) = \frac{d}{dt} \int_0^t A(t - \tau)E(\tau)\, d\tau\, ^{57} \qquad [6.17]$$

Carson extended the range of the Heaviside operational calculus by making use of Borel's theorem of real convolution.[58] An integral equation of the form

$$x(t) + m \int_0^t K(t - T)x(T)\, dT = x_0(t) \qquad (6.18)$$

appeared in a paper by Poisson in 1826, and Duhamel (1833), Boltzmann (1874), John Hopkinson (1887), and Chebyshev (1887) all made use of convolution.[59] Using the extended Heaviside calculus, Carson investigated between 1917 and 1926 many electrical transient phenomena.[60]

The techniques used were largely based on the use of the Fourier integral theorem, which was developed into a standard transform technique by Campbell and Foster, who in 1931 published an extensive table of transform pairs.[61]

Although simple to apply, the Heaviside operational calculus requires all initial conditions to be zero and lacks any precise conditions for assessing its validity — approximate conditions can be obtained by analogy with the Laplace-transform method. The use of the Fourier transform removes these two restrictions, but its general application is restricted because of the inherent restrictions on transformable functions. Bateman in a paper written in 1910 suggested the use of the Laplace transform and in 1921 wrote to Carson recommending its use.[62]

Although there was gradual acceptance of the Laplace transform during the 1920s the emphasis was on the Heaviside operational calculus and the application of Fourier transforms. Books such as Carslaw's *Introduction to the theory of the fourier integral*[63] published in 1921 and Carson's *Electric circuit theory and the operational calculus*[64] were followed by those of Jeffreys (1927), Cohen (1928), Woodruff (1928), Berg (1929), Bush (1929), Campbell and Foster (1931) and others.[65] It was not until the late 1930s and early 1940s that the more general Laplace-transform approach became widely known, with the publication in Germany in 1937 of a book by Doetsch,[66] followed by the books by Carslaw and Jaeger (1941), Churchill (1941) and Gardner and Barnes (1942).[67]

To the practical engineer these changes meant little. Faced with the problem of the solution of differential equations he, like Heaviside, turned the algebraic handle, trusting to his understanding of the physical system to warn him of any mathematical difficulties due to problems with initial conditions, and so on. However, the mathematical developments had changed the climate of opinion; communications engineers in the late 1920s were beginning to talk in terms of spectra and of bandwidth, and the more mathematically minded were venturing into the realms of complex-function theory. Thus the foundations for the development of control theory and control-systems design techniques were laid.

References and notes

1 DERRY, T. K., and WILLIAMS, T. I.: *A short history of technology* (Oxford University Press, 1960), pp. 621–622

2 THOMSON, W.: 'On the theory of the electric telegraph', *Philosophical Magazine*, 1856, **11**, pp. 144–160

3 CROWTHER, J. G.: *American Men of Science – 2*, (Penguin Books, Harmondsworth, 1944), p. 63

4 The notebooks are now in the possession of the Institution of Electrical Engineers; the lines are quoted from Sumpner, W. E.: 'The work of Oliver Heaviside', *J. IEE*, 1932, 71, p. 838. See also an article by E. T. Whittaker, which has been reprinted in MOORE, D. H.: *Heaviside operational calculus: an elementary foundation* (Elsevier, New York, 1971)

5 FLEMING, J. A.: *Scientific research and electrical engineering* (Constable, London, 1927), p. 144. Michael Idvorsky Pupin, born in Yugoslavia, arrived in the USA in 1874, destitute; he learnt English and won a scholarship to Columbia University in 1879, graduating in 1883. After graduating he studied at Cambridge University, where he was tutored by Routh, before going to Berlin to study under Helmholtz. He taught at Columbia University from 1889 to 1931.

6 COLPITTS, E. H.: 'Introduction', *Collected papers of George Ashley Campbell* (American Telephone and Telegraph Company, New York, 1937), pp. 2–3

7 BODE, H. W.: 'Feedback – the history of an idea', *Proceeedings of the Symposium on Active Networks and Feedback Systems* (Polytechnic Press, Brooklyn, 1960), p. 4; reprinted in BELLMAN, R., and KALABA, R.: *Mathematical trends in control theory* (Dover, New York, 1964), p. 110

8 GILL, F.: 'Notes on the howling telephone', *J. IEE* 1901, **31**; it is a little surprising that this discovery was not followed up, as shortly afterwards radio engineers were beginning to look for oscillator circuits

9 CAMPBELL, G. A., 1912; quoted from COLPITTS: *op. cit.*, p. 7

10 BAKER, E. C.: *Sir William Preece F.R.S. Victorian engineer extraordinary* (Hutchinson, London, 1976), chap. 28, pp. 266–280; gives an account of the assistance (and lack of assistance) given by the Post Office and Government to Marconi and to the Marconi Wireless & Telegraph Company during the early years. Sir William Preece (1834–1913) was for many years Chief Engineer to the Post Office. A. A. Campbell Swinton (1863–1930) educated at Edinburgh and the Elswick Works of Sir W. G. Armstrong was a member of the three senior engineering institutions and became Director of BSIRA

11 BAKER, W. J.: *A history of the Marconi Company* (Methuen, London, 1970), pp. 106–107

12 There was extensive litigation between de Forest and Edwin H. Armstrong regarding priority over the invention of the Audion. In 1934 the United States Supreme Court decided in favour of de Forest; the electrical engineering profession, however, has always considered Armstrong as having the strongest claim. For a brief survey of the dispute, see TUCKER, D. G.: 'The history of positive feedback: the oscillating Audion, the regenerative receiver, and other applications up to around 1923', *Radio and Electron. Eng.*, 1972, **42**, pp. 69–80

13 de FOREST, L.: 'The Audion', *Trans. AIEE*, 1906, **25**, p. 735

14 Harold D. Arnold (1883–1935) was Director of Research, Western Electric, from 1917–1924, and of the Bell Telephone Laboratories from 1925 until his death in 1933

15 Irving Langmuir (1881–1957), a physicist of distinction, worked in the General Electric Research Laboratories between 1906 and 1909; from 1909 until his death he was a consultant to the General Electric Company

16 E. W. F. Alexanderson joined Steinmetz at General Electric in 1902 and served the company for 50 years

17 BRITTAIN, J. E.: 'C. P. Steinmetz and E. F. W. Alexanderson, creative engineering and a corporate setting', *Proc. IEEE*, 1976, **64**, p. 1415

18 US Patent 1 113 149, filed 29 October 1913, issued 6 October 1914; extracts quoted from TUCKER, D. G.: *op. cit.*, p. 72

19 ARMSTRONG, E. H.: 'Some recent developments in the audion receivers', *Proc. Inst. Radio Eng.*, 1915, **3**, p. 215

20 Quoted from TUCKER: *op. cit.*, p. 73

21 *ibid.*, p. 73

22 *ibid.*, pp. 74–75

23 British Patent 13 636, 12 June 1913, issued 11 June 1914; passage quoted from TUCKER, D. G.: *op. cit.*, p. 74

24 *ibid.*, p. 74

25 *ibid.*, pp. 76–78, for details

26 HAZELTINE, L. A.: 'Oscillating audion circuits', *Proc. Inst. Radio Eng.*, 1918, **6**, p. 63

27 VAN DER POL, B.: 'A theory of the amplitude of free and forced triode vibrations', *Radio Review*, 1920, **1**, p. 701

28 APPLETON, E. V., and VAN DER POL, B.: 'On a type of oscillation-hysteresis in a simple triode generator', *Philosophical Magazine*, 1922, **43**, p. 177, and APPLETON, E. V.: 'Automatic synchronization of triode oscillators', *Proceedings of the Cambridge Philosophical Society*, 1922–1923, **21**, p. 231

29 BODE, H. W.: 'Feedback – the history of an idea', *Proceedings of the Symposium on Active Networks and Feedback Systems*, 1960, p. 4. Reprinted in Bellman, R., and Kalaba, R.: *Selected papers on mathematical trends in control theory* (Dover, New York, 1964), pp. 106–123

30 COLPITTS: *op. cit.*, p. 4

31 FRIIS, H. T., and JENSEN, A. G.: 'High frequency amplifiers', *Bell Syst. Tech. J.*, 1924, **3**, pp. 181–205

32 An extensive investigation of the form and behaviour of such networks had been carried out by Campbell and Foster, of the Bell Telephone Laboratories in connection with repeater stations; see CAMPBELL, G. A., and FOSTER, R. M.: 'Maximum output networks for telephone substations and repeater circuits', *Trans. AIEE*, 1920, **39**, pp. 231–280

33 BODE: *op. cit.*, pp. 6–7

34 US Patent, 2102671, 1927

35 BLACK, H. W.: 'Stabilized feedback amplifiers', *Bell Syst. Tech. J.*, 1934, **13**, pp. 1–18

36 *ibid.*, pp. 1–2

37 *ibid.*, p. 2, footnote

38 *ibid.*, p. 3

39 *ibid.*, p. 5

40 *ibid.*, p. 2

41 *ibid.*, p. 11

42 BODE, H. W.: 'Relations between attenuation and phase in feedback amplifier design', *Bell Syst. Tech. J.*, 1942, **19**, p. 431

43 NYQUIST, H.: 'Regeneration theory', *Bell Syst. Tech. J.*, 1932, **11**, pp. 126–147

44 HAZEN, H. L.: 'Theory of servo-mechanisms', *J. Franklin Inst.*, 1934, **218**, pp. 279–233

45 IVANOFF, A.: 'Theoretical foundations of the automatic regulation of temperature', *J. Inst. Fuel*, 1934, **7**, pp. 117–130

46 Kennelly's ideas on the analysis of alternating-current circuits were crystallised in his paper 'Impedance', *Trans. AIEE*, 1893, **10**

47 Steinmetz presented his method of analysing electrical circuits at the International Electrical Congress in Chicago in 1893, although details of the method were not published until four years later, and it was many years before electrical engineers became conversant with his ideas (*Dictionary of American Biography*)

48 THOMSON: *op. cit*

49 SUMPNER: *op. cit.*, p. 845

50 STEPHENS, E.: *The elementary theory of operational mathematics* (McGraw-Hill, New York, 1937), p. 268

51 BROMWICH, T. J. I'A.: 'Normal coordinates in dynamical systems', *Proceedings of the London Mathematical Society*, 1916, **15**, pp. 401–448; see also CARSLOW, H. S., and JAEGER, J. C,: *Operational methods in applied mathematics* (Oxford University Press, 1941), pp. xiii–xiv

52 Similar ideas for the solution of differential equations were being developed by Doetsch, Giorgi, Wagner, Bateman and others; see, for example, GARDNER, M. F., and BARNES, J. L.: *Transients in linear systems*, (Wiley, New York, 1942), p. 362, but Bromwich's work had the most influence on the application of the techniques in the analysis of electrical networks

53 CARSON, J. R.: 'General expansion theorem for the transient oscillations of a connected system', *Physical Review*, 1917, **10**, pp. 217–225

54 FRY, T. C.: 'The solution of circuit problems', *Physical Review*, 1919, **14**, pp. 115–136

55 CARSON, J. R.: 'Theory of transient oscillations of electrical networks and transmission systems', *Trans. AIEE*, 1919, **38**, pp. 407–489

56 CARSON, J. R.: 'Electric circuit theory and the operational calculus', *Bell Syst. Tech. J.*, 1925, **4**, p. 629

57 *ibid.*, p. 702

58 The real convolution theorem is usually named after Borel, but he was not the first to give a rigorous proof; see GARDNER and BARNES: *op. cit.*, p. 364

59 *ibid.*, p. 364

60 This work was presented in a series of 15 lectures given at the Moore School of Engineering, University of Pennsylvania, in the spring of 1925. The lectures were published in the *Bell System Technical Journal*, 1925, **4**, pp. 685–761; 1926, **5**, pp. 50–95, 336–384, and then issued as a book, *Electric circuit theory and the operational calculus* (McGraw–Hill, New York, 1926)

61 CAMPBELL, G. A., and FOSTER, R. M.: *Fourier integrals for practical applications* (Bell Telephone System, New York, 1931)

62 BATEMAN, H.: 'The control of an elastic fluid', *Bull. Am. Math. Soc.*, 1945, **51**, p. 644; reprinted in BELLMAN, R., and KALABA, R.: *Mathematical trends in control theory* (Dover, New York, 1964), pp. 18–64

63 CARSLAW, H. S.: *Introduction to the theory of Fourier's series and integrals* (Macmillan, New York, 1921)

64 CARSON, J. R.: *Electric circuit theory and operational calculus* (McGraw–Hill, New York, 1926)

65 JEFFREYS, H.: *Operational methods in mathematical physics* (Cambridge University Press, 1927); COHEN, L. I.: *Heaviside's electrical circuit theory* (McGraw–Hill, New York, 1928); WOODRUFF, L. F.: *Principles of electric power transmission and distribution* (Wiley, New York, 1928); BERG, E. J.: *Heaviside's operational calculus* (McGraw–Hill, New York, 1929); BUSH, V.: *Operational circuit analysis* (Wiley, New York, 1929); CAMPBELL, G. A., and FOSTER, R. M.: *Fourier integrals for practical applications* (Bell Telephone System, New York, 1931), Bell System Monograph B-584, 1931

66 DOETSCH, G.: *Theorie und Anwendung der Laplace-Transformation* (Springer, Berlin, 1937)

67 CARSLAW, and JAEGER: *op. cit.*; CHURCHILL, R. V.: *Operational mathematics* (McGraw–Hill, New York, 1944); GARDNER and BARNES: *op. cit*

Short bibliography of books and articles on the history of control engineering

ANDRONOV, A. A., and VOSNESENSKII, I. N.: 'The work of J. C. Maxwell, A. I. Vyshnegradskii, and A. Stodola in the theory of machine control' *in A. A. Andronov: Sobranie Trudov* (Moscow, Izdat, ANSSSR, 1956) (in Russian)

BATEMAN, H.: 'The control of an elastic fluid', *Bull. Am. Math. Soc.*, 1945, **51**, pp. 601–646. (The 17th Josiah Willard Gibbs lecture, an important survey of the state of control theory and its relationship to engineering practice)

BENNETT, S.: 'The search for "uniform and equable motion": a study of the early methods of control of the steam engine', *Int. J. Control*, 1975, **21**, pp. 113–147

BENNETT, S.: 'The emergence of a discipline: automatic control 1940–1960', *Automatica*, 1976, **12**, pp. 113–121

BENNETT, S.: 'A note on the early development of control', *J. Dynamic Systems, Measurement & Control, Trans. ASME*, 1977, **99**, pp. 211–213

COALES, J. F.: 'Historical and scientific background of automation',*Engineering*, 1956, **182**, pp. 363–370

CONWAY, H. G.: 'Origins of mechanical servo-mechanisms', *Trans. Newcomen Soc.*, 1953–55, **29**, pp. 55–75

FULLER, A. T. (ed): *Stability of motion* (Taylor & Francis, London, 1975). (This is a reprint of E. J. Routh's essay 'Stability of a given state of motion', originally published in 1877. It also includes other papers by Routh and by Clifford, Sturm and Böcher. Fuller provides an introductory essay giving some of the background to Routh's work)

FULLER, A. T.: 'The early development of control theory – I', *J. Dynamic Systems, Measurement, & Control, Trans. ASME*, 1976, **98**, pp. 224–235 (Surveys early governors which make use of integral action; main body of paper deals with Maxwell's work)

FULLER, A. T.: 'The early development of control theory – II', *J. Dynamic Systems, Measurement, & Control, Trans. ASME*, 1976, **98**, pp. 224–235

FULLER, A. T.: 'Edward John Routh', *Int. J. Control*, 1977, **26**, pp. 169–173

GIBBS, F. W.: 'The furnaces and thermometers of Cornelius Drebbel', *Ann. Sci.*, 1950, **6**, pp. 32–43

HORT, W.: 'Die Entwicklung des Problems der stetigen Kraftmaschinenregelung nebst einem Versuch der Theorie unstetiger Regelungsvorgänge', *Z. Mathematik & Physik*, 1904, **50**, pp. 233–279

HUGHES, T. P.: *Elmer Sperry: inventor and engineer* (Johns Hopkins Press, 1971, Baltimore)

KHRAMOI, A. V.: *History of automation in Russia before 1917* (Jerusalem, 1969)

MAYR, O.: *The origins of feedback control* (MIT Press, Cambridge, Mass. 1970). (The major English language work on feedback control prior to 1800. Originally published in German as *Zur Frühgeschichte der Technischen Regelungen* (Oldenbourg, 1969, München)

MAYR, O.: *Feedback mechanisms in the historical collections of the National Museum of History and Technology*, Smithsonian Studies in History and Technology 12 (Smithsonian Institute Press, Washington, 1971)

MAYR, O.: 'Adam Smith and the concept of the feedback system', *Technology & Culture*, 1971, **12**, pp. 1–22

MAYR, O.: 'Victorian physicists and speed regulation: an encounter between science and technology', *Notes and Records of the Royal Society of London*, 1971, **26**, pp. 205–228. (Deals with the work on governors of Airy, Siemens, Foucault, Kelvin, Maxwell and Gibbs)

MAYR, O.: 'James Clerk Maxwell and the origins of cybernetics', *Isis*, 1971, **62**, pp. 425–444

PRIME, H. A.: 'The History of automatic control', presented at the IEE Conference on the history of electrical engineering, 1973

RAMSEY, A. R. J.: 'The thermostat or heat governor: an outline of its history', *Trans. Newcomen Soc.*, 1945–47, **25**, pp. 53–72

RÖRENTROP, K.: *Entwicklung der modernen Regelungstechnik* (R. Oldenbourg, 1971, München)

SIEMASZKO, Z. A.: 'Control before the 20th century', *Control & Instrum.*, July 1979, **1**, pp. 65–67; Aug. 1969, **1**, pp. 45–47

TUCKER, D. G.: 'The history of positive feedback: the oscillating audion, the regenerative receiver, and other applications up to around 1923', *Radio & Electron. Eng.*, 1972, **42**, pp. 69–80

For the governor shown in Fig. 2.14, assume

$$AB = BC = a; \text{hence } h = 2a\cos\alpha.$$

Taking moments about A for half the governor,

$$m\omega^2 b \sin\alpha.b\cos\alpha = mgb\sin\alpha + Mga\sin\alpha,$$

which gives

$$b\cos\alpha = \frac{g}{\omega^2} + \frac{Mg}{m\omega^2}\frac{a}{b}$$

Substituting for $\cos\alpha$ in terms of h and a gives

$$b\frac{h}{2a} = \frac{g}{\omega^2} + \frac{Mg}{m\omega^2}\frac{a}{b}$$

Note that, if $a = b$ and $M = 0$, $h/2 = g/2$, where $g/\omega^2 = $ height of simple conical pendulum.

The capacity of a governor is defined as $C = $ restoring force × distance moved:

$$\tfrac{1}{2}C = (m + Ma/b)g(h_1 - h_2)$$

$$h_1 = \frac{2a}{b}\frac{g}{\omega_1^2}\left(\frac{m + Ma/b}{m}\right)$$

$$h_2 = \frac{2a}{b}\frac{g}{\omega_2^2}\left(\frac{m + Ma/b}{m}\right)$$

Hence

$$\tfrac{1}{2}C = (m + Ma/b)^2\frac{2a}{b}\frac{g^2}{m}\left(\frac{1}{\omega_1^2} - \frac{1}{\omega_2^2}\right)$$

The effect of loading on capacity is shown in the graph.

Detent due to friction: With friction force dF,

$$h = \frac{2a}{b} \frac{g}{\omega^2} \left| \frac{m + (F/g)\,sgn(\delta\omega)a/b}{m} \right|$$

which gives, for unloaded governor,

$$\frac{\omega_n - \omega_l}{\omega} = \frac{F}{gm}$$

and, for loaded governor,

$$\frac{\omega_n - \omega_l}{\omega} = \frac{F}{(m + M)g}$$

Index

Printed in the USA
CPSIA information can be obtained
at www.ICGtesting.com
JSHW011519221024
72172JS00008B/67